# BIRDS NORTHUMBRIA

The 1992 Bird Report for Northumberland,
Newcastle and North Tyneside

by

DAVID JARDINE, ALAN JOHNSTON, IAN KERR, DAVID McKEOWN
and NICHOLAS ROSSITER

*Illustrations by*
ANDREW BOOTH
ALAN HART
STEWART SEXTON
JOHN STEELE

*Published by*
THE NORTHUMBERLAND AND TYNESIDE BIRD CLUB
Registered Charity No. 517641

© Copyright, Northumberland and Tyneside Bird Club 1993

# NORTHUMBERLAND AND TYNESIDE BIRD CLUB OFFICIALS

*Chairman*
ALAN JOHNSTON
39 Cranwell Drive
Woodlands Park
Wideopen
Newcastle upon Tyne
NE13 6AS

*Honorary Secretary*
ANDREW BRUNT
7 Highmoor
Kirkhill
Morpeth
Northumberland
NE61 2AL

*Honorary Treasurer*
LINDSAY McDOUGALL
45 Pendleton Drive
Northburn Chase
Cramlington
Northumberland
NE23 9TU

*County Recorder*
NICHOLAS ROSSITER
West Barn
Lee Grange
Hexham
Northumberland
NE46 1SX

*Honorary Auditor*
BRIAN ARMSTRONG I.P.F.A.

Alexander Bankier, Keith Brooks, Martin Davison, David Jardine and Geoff Linkleter also served on the Club Committee during the year.

Entries in this report have been scrutinised by the County Records Committee consisting of Colin Bradshaw, Martin Davison, Brian Little, Nicholas Rossiter (chairman) and James Steele. Michael Hodgson also served on the committee during the year. The appearance of a record indicates its acceptance by this committee but not necessarily by the British Birds Rarities Committee, although any decision by the latter will be referred to in the text.

ISBN 0-9522039-0-1

# CONTENTS

Introduction ................................................. 4

Monthly Summary ............................................. 5

Classified List ................................................ 8

List of Contributors ......................................... 90

Additions and Corrections to Previous Reports ................ 91

Decisions Pending ........................................... 94

Exotic Species ............................................... 95

The Nightjar Breeding Survey 1992 ........................... 96

The Shelduck Breeding Survey 1992 ........................... 99

New Species for the County     Pied-billed Grebe ............. 101

                                   Semipalmated Sandpiper ....... 102

                                   Swinhoe's Storm-petrel ........ 103

# INTRODUCTION

The entries in this report cover all species found during 1992 in our recording area which comprises the county of Northumberland and the districts of Newcastle and North Tyneside.

The Classified List follows the sequence compiled by Professor K. H. Voous in 1977 with the status of each species categorised under one of six headings:-

| | |
|---|---|
| Extremely rare | Up to ten records in total |
| Rare | One–ten occurrences annually |
| Uncommon | 11–100 occurrences annually |
| Well-represented | 101–1,000 occurrences annually |
| Common | 1,001–10,000 occurrences annually |
| Abundant | Over 10,000 occurrences annually |

Five authors have compiled the species accounts with responsibility as follows:-

| | |
|---|---|
| Divers–Sawbills | Alan Johnston |
| Raptors–Skuas | Ian Kerr |
| Gulls–Chats | Nicholas Rossiter |
| Thrushes–Shrikes | David McKeown |
| Crows–Buntings | David Jardine |

Ian Kerr provided the Introduction and Monthly Summary and Nicholas Rossiter the Amendments and Corrections to Previous Reports, Pending Decisions and details of Exotic Species.

1992 provided two new species for the county list, Pied-billed Grebe and Semipalmated Sandpiper, while the fascinating mystery of the dark-rumped petrels caught at Tynemouth between 1989–1992 was finally solved when they were identified as Swinhoe's Storm-petrels. Another highlight was the first breeding record of Montagu's Harriers for 26 years.

The year also provided an impressive array of species in the extremely rare and rare categories as an examination of the classified list will show.

The knowledge of the County Records Committee was again exercised to the full. However, once again, some records were reluctantly rejected because, despite repeated requests, observers failed to supply sufficient descriptions and, in some cases, none at all.

The Classified List has been drawn up from over 12,000 record cards submitted by members and others and from data supplied for the Farne Islands by John Walton of the National Trust and for Coquet Island by the Royal Society for the Protection of Birds.

The authors thank all contributors, our artists, those who have provided special papers on breeding surveys and new species and Bryan Galloway and Tom Cadwallender for the ringing data.

All records for the 1993 report, with descriptions where required or requested, should be in the hands of the County Recorder by 31st January 1994 to enable prompt publication.

# MONTHLY SUMMARY

1992 began with wet and windy weather but the final two weeks of January were dominated by high pressure leading to sharp frosts which left many ponds frozen. Typically, mid-winter wildfowl, including Whooper and Bewick's swans, Brent, White-fronted and Greylag geese were an attraction, as were several Smew. Unseasonal waders were represented by Ruff, Black-tailed Godwit and Spotted Redshank. Among smaller species, Waxwings were prominent, Twite were present and there was an encouraging number of Corn Buntings. As usual, a few wintering Blackcaps and Chiffchaffs were found, mainly in gardens providing both shelter and feeding. A Black Redstart was very unusual. However, all were eclipsed by the discovery of a splendid male Pine Bunting which provided excellent viewing for many hundreds of observers. Harbours held Glaucous, Iceland and Mediterranean gulls.

The very cold conditions continued into early February before a thaw set in only to be followed by snow. However, by the end of the month conditions were mild, prompting many species into song. The Pine Bunting remained a very big attraction. Waxwings continued to entertain in many areas and the Black Redstart stayed. There was a light influx of Greenland White-fronted Geese but the pick of wildfowl was a drake Green-winged Teal. Buzzards and wintering Hen Harriers were noted in several areas. By the end of the month some waders were moving back into breeding areas and there were the usual early passages involving Common Gulls. The various 'white' gulls remained.

March proved fickle with periods of mild and spring-like conditions followed by some strong winds but there were no extremes of weather or significant snow or heavy rain. Offshore, Long-tailed Ducks were present in good numbers, the Greenland White-fronted Geese remained and a light passage of Pintails was evident. Much more unusual was an inland Red-throated Diver. The usual early summer visitors began to appear: Wheatear on 14th, a Sandwich Tern on 18th and Sand Martins from 23rd. Water Pipits put in an appearance, there was the start of a very good passage of White Wagtails and more Waxwings were passing through the area. The pick of the March raptors was a Golden Eagle.

April began with a deluge causing widespread flooding and swollen rivers with some areas affected by up to 15 cms of rain in two days. On higher ground these conditions led to the heaviest snowfalls of the winter leaving many deep drifts. However, conditions quickly improved and late in the month winds from S prompted the arrival of many summer visitors. The first Marsh Harriers and Ospreys passed through, Garganey appeared on several ponds while the arrival of summering waders were represented by Little Ringed Plovers, Ruffs and Common Sandpipers. Many Waxwings remained as more summer visitors flooded in with House Martins on 7th, Cuckoo and Ring Ousel on 8th, Common Tern and Garden Warbler on 9th and Swallow on 10th. Yellow Wagtails were first seen on 12th, Swifts and Grasshopper Warblers on 21st with Common and Lesser whitethroats, Redstarts, Whinchats, Wood and Sedge warblers all following by the end of the month. Rarities included a Hoopoe, Firecrest and Shore Lark. Large numbers of passage White Wagtails were found.

Early May was hot and sunny with a record-breaking spring temperature of 27C in Newcastle during the first week, although sea breezes kept conditions cooler on the coast. In these excellent conditions spring arrival continued apace. More Marsh Harriers and Ospreys featured while spring Hobbys passed through and a Honey Buzzard was seen. Heavy arrival of terns occurred, including the return of the regular Lesser Crested Tern. May is usually a month for spring rarities and 1992 was no exception. Sought-after species included Little Egret, Spoonbill, rarer races of Yellow Wagtail as well as Subalpine Warbler, Bluethroat, Golden Oriole, Red-backed Shrike and, more surprisingly, a Woodchat Shrike and a Rustic Bunting were reported.

Early June continued to be hot, dry and sunny with winds predominantly from the E producing a fall of late spring migrants along the coast. These included Marsh and Icterine warblers, more Red-backed Shrikes, Scarlet Rosefinches and another Golden Oriole. The county's first Semipalmated Sandpiper was found and a drake Surf Scoter was joined by a drake King Eider. Among breeding species the rapid spread of Ruddy Ducks continued, survey workers found many more Nightjars than had been previously suspected and Black-necked Grebes spread to new waters. Among the rarer raptors, Buzzards and Hen Harriers had a good breeding season with reasonable results also reported for Peregrines and Merlins.

Temperatures dropped in early July with some unseasonable cool spells and some showery and windy periods. There was the now annual return of a dark-rumped petrel, eventually to be identified as Swinhoe's Petrel. Storm Petrels were offshore and among the rare breeding species Little Ringed Plovers raised young. Rarities continued to be a big attraction with a Great White Egret, Honey Buzzard and Spoonbill being found but sea-watching was disappointing with few shearwaters in evidence. By the second half of the month return wader passage was beginning with Sanderlings, Little Stints, Whimbrels and Curlew Sandpipers.

August was a month of showers, below average temperatures, winds coming mainly from W and SW with some periods of heavy rain in the closing days. More evidence of the spread of breeding Buzzards was found although these were eclipsed by the first confirmed breeding success for Montagu's Harrier for 26 years. Elsewhere the main interest lay on the coast as wader passage quickly gathered pace with the start of a big build up, particularly of Ruffs. However, a Pectoral Sandpiper was the pick of the waders. Another Sabine's Gull was found, skuas were passing in small numbers and Black Terns began to appear. Also found was Corncrake and unseasonal Glaucous and Iceland gulls.

September was a month of mixed weather with warm and sunny periods alternating with some showery and windy spells. As usual at this time of year rarities were well to the fore with another Sabine's Gull and several Long-tailed Skuas. Unsettled conditions late in the month produced a run of passerine rarities including Bluethroat, Red-breasted Flycatcher, Arctic, Marsh and Yellow-browed warblers, Little Bunting and Hoopoe. A Yellow-breasted Bunting was a huge attraction. Wader passage continued with high numbers of Ruffs and a Temminck's Stint was found. Wildfowl were on the move with passage of Barnacle and Canada geese while Brent Geese were also arriving in good numbers.

Temperatures in October were well below normal as a series of depressions moved from the N and E. These conditions produced an incredible and unprecedented passage of Pomarine Skuas which, for many observers, was the highlight of the year. The winds also prompted other sea passage with good numbers of divers and wildfowl, particularly Velvet Scoters and another Surf Scoter appeared. Wildfowl numbers continued to build up and the year's second Green-winged Teal was found. Despite these apparently promising conditions there were none of the expected big autumn falls of thrushes, Goldcrests and Robins. However, those seeking rarities were not disappointed. Among the rare species of the month was a Bonelli's Warbler, more Yellow-browed Warblers, Rustic and Ortolan buntings and a Wryneck.

After the usual excitement of October, November usually proves to be an anti-climax and this was the case. It was a dull and rather wet month, was generally cold and some snow fell on the hills. Wintering geese continued to build up as did Whooper Swans and several Smew arrived to grace local waters. Raptors were well reported with the pick being a Rough-legged Buzzard and Hen Harriers were settling into wintering hunting areas. As usual for November, some summer visitors lingered with late Swallows and Ring Ousels still present as typical wintering species, including Twite, were moving into the region.

The first half of December was mild and wet but then conditions changed dramatically with a settled high pressure system producing bright but very cold days and night temperatures dropping as low as –9C. Inland waters quickly became icebound, as did coastal ponds towards the end of the month. However, these Arctic conditions did not deter hundreds of observers who enjoyed the county's first Pied-billed Grebe which remained into 1993. Smew were on several waters, September's Green-winged Teal remained and, despite the very inhospitable conditions, wintering Green Sandpipers were discovered. Among smaller species, a Shore Lark was found, there was an extremely late Swallow, Great Grey Shrikes were an attraction and wintering Blackcaps and Chiffchaffs remained in the gardens of some luckier observers.

# CLASSIFIED LIST

**Red-throated Diver,** *Gavia stellata*
A common passage and winter visitor.
During the first quarter the usual favoured feeding areas held the main concentrations. The area off Stag Rocks, Bamburgh, held 20 in January although peak numbers occurred in March when 48 were off Ross and 28 in Druridge Bay. Smaller groups were noted off Seaton Sluice and Newbiggin.
Light passage N towards breeding grounds occurred from late March and then one-two summering individuals remained and were reported between June-August.
Return passage began in September with 156 S and 30 N at Seaton Sluice during the month. Numbers peaked in October when 100 gathered in Druridge Bay and observers at Seaton Sluice counted 215 N and 109 S. Smaller movements were seen from Bamburgh and Holy Island.
Freshwater records are unusual and involved singles at Ladyburn Lake in February and October, at Derwent in March and Catcleugh in April

**Black-throated Diver,** *Gavia arctica*
An uncommon passage and winter visitor.
Between January-March two-three were in Druridge Bay and one-two in Beadnell Bay and off Seahouses and Bamburgh. During May two off Ross and a single off the Long Nanny Burn probably involved birds passing N.
The species was then absent until 26th July when a single was in Druridge Bay. One-two continued to be noted in this area until September when on 18th return passage was marked by three S at Seaton Sluice and one off Newbiggin.
Sightings increased in October with four off Ross and Hauxley and one-three seen from Seaton Sluice, Newbiggin, Druridge, Seahouses, the Farne Islands and Holy Island.
During November-December wintering singles were noted at five localities.

**Great Northern Diver,** *Gavia immer*
An uncommon passage and winter visitor.
Two were off Stag Rocks, Bamburgh, on 4th January while elsewhere during the first quarter singles occurred in Druridge Bay, again off Bamburgh and at Scremerston.
Passage was evident with two N at Newbiggin on 31st March. From April-September singles were off Newbiggin, Cresswell, Seahouses, the Farne Islands and Cocklawburn.
Return passage occurred in October with three N and one S at Seaton Sluice between 9th-20th and one-two were also seen from Newbiggin, Hauxley, the Farne Islands, Ross and Holy Island. During November-December the only reports were of singles in the favoured wintering area between Bamburgh-Holy Island.

**Pied-billed Grebe,** *Podilymbus podiceps*
An extremely rare visitor.
An individual found at Druridge Pools on 26th December remained until at least 19th January 1993. It was the first county record and a special

paper documenting its discovery appears later in this report. The record is still under consideration by the British Birds Rarities Committee.

PIED-BILLED GREBE – a new species for the region

**Little Grebe,** *Tachybaptus ruficollis*
An uncommon breeding species, well-represented as a passage visitor.
　　The largest early gatherings involved 16 at Cresswell Pond in January-February when five were at Hauxley Reserve. One-four frequented eight other ponds and rivers during the period.
　　A return to breeding areas occurred in March with sightings at ten localities. During April nine were at Warkworth Lane Pond and one-six at a further 18 sites.
　　The first young were found at Arcot Pond on 23rd May and five broods were reported from this locality by the end of the season. Successful breeding also occurred at Farnley (four pairs), Hauxley Reserve, Howick and Linton ponds and Grindon Lough (all two pairs) and single broods were noted at Throckley, Wallsend Swallow and Warkworth Lane ponds and at Big Waters.
　　Post-breeding gatherings included 17 at Bothal Pond in August, 16 at Hauxley Reserve in September while during October 17 were on Cresswell Pond. Sightings were scarce in November-December with reports from only four localities, 14 at Cresswell Pond being the largest group.

**Great Crested Grebe,** *Podiceps cristatus*
A rare breeder and an uncommon passage and winter species.
　　Up to five were in Druridge Bay in January-February and during the same period three visited Holywell Pond, Big Waters and Whittle Dene.
　　During late March three pairs were at Holywell Pond and one-four birds were seen at a further nine localities where display was noted. Successful breeding followed at Holywell Pond (three broods), Whittle Dene (three pairs present) and single broods were found at Big Waters and Farnley.
　　During September eight were at Colt Crag, five at Whittle Dene and one-

three at five other waters. During the final quarter 11 were in Druridge Bay in October when one-three were at a further five coastal sites. In November singles remained inland at Big-Waters and Hallington.

**Red-necked Grebe,** *Podiceps grisegena*
An uncommon passage and winter visitor.
　　The main wintering area between Holy Island-Bamburgh held an early peak of eight in January-February while elsewhere the only reports involved singles at Newton and in Blyth Bay.
　　During April three in summer plumage were off Bamburgh and another was in Holy Island harbour. Summering was suggested by singles off Bamburgh in late May and mid-July.
　　The first indication of return passage was on 29th August when two were seen off Ross, followed in September by one-two at four localities. Numbers increased in October with six off Holy Island and one-two reported from seven localities, including Ladyburn Lake where a single remained from 18th-28th.
　　A loose gathering of 15 off Ross on 21st November was the highest count in the remainder of the year.

**Slavonian Grebe,** *Podiceps auritus*
An uncommon passage and winter visitor.
　　The favoured area between Holy Island-Bamburgh held an early peak of 30 off Ross on 8th March. They were characteristically scarce elsewhere with the only other reports involving one-two at N Blyth in February.
　　The species was absent from April until 5th September when one flew N at Newbiggin. During the final quarter birds were again settled in the usual wintering area with 20 off Ross in November. Sightings of eight off Stag Rocks and six at Holy Island probably involved some of these birds. Singles were on Ladyburn Lake in October and Cresswell Pond in November.

**Black-necked Grebe,** *Podiceps nigricollis*
A rare breeder and an uncommon passage and winter visitor.
　　Spring records involved one-two at Cresswell, Warkworth Lane and Holywell ponds, Druridge Pools and Caistron between April-June.
　　Seven returned to the regular breeding site by 17th April, rising to a peak of 24 in June. After the total failure of this site in 1991 there was only a slight improvement with four pairs going on to raise one youngster each.
　　At a second water where successful breeding occurred in 1991 three pairs were present in April-May but no young were seen. However, colonisation occurred at two further waters. Three pairs raised a total of four young at one site while at the other a pair hatched four young but two were predated by Lesser Black-backed Gulls, *Larus fuscus graellsi*.
　　Dispersal from breeding areas resulted in a single adult at Ladyburn Lake on 25th-26th July, a juvenile at Holywell Pond between 28th July-9th August and singles at Whittle Dene in August and Cresswell Pond on 26th September. The final sighting was a single with Slavonian Grebes, *Podiceps auritus*, off Holy Island on 31st October.

**Fulmar,** *Fulmarus glacialis*
A well-represented breeding species, abundant on passage.

Large numbers frequented at least seven breeding cliffs and quarries during the first quarter.

Spring passage N occurred in April with the heaviest movement involving 500 per hour at Seaton Sluice on 15th when most local breeders were already settled on nesting sites.

During the season, 223 pairs bred on the Farne Islands (260 in 1991) and 57 pairs on Coquet Island (34 in 1991). Other favoured cliffs also held large concentrations with 356 pairs between Berwick and the Border, 78 pairs at Tynemouth, 66 at Cullernose Point, 43 at Howick, 36 at Dunstanburgh, 28 at Coves Bay, Holy Island and 28 pairs at Old Hartley. Once again five pairs used ledges at Blyth Power Station and there was one pair on Berwick road bridge.

The heaviest autumn movement was of only 100 per hour N at Craster on 6th September before most birds moved away for the moulting period. By mid-November 300 were back around the Farne Islands with smaller numbers again frequenting mainland cliffs and quarries.

'Blue' phase individuals were particularly scarce with the only sighting involving a single N at Seaton Sluice on 5th September.

As usual a few birds wandered inland with the furthest W being a single over Chapel House, Newcastle, on 22nd June.

**Sooty Shearwater,** *Puffinus griseus*
A well-represented passage visitor.

The first typical mid-summer passage individuals were seen in July. One moved N at Hauxley on 5th and another followed at the same locality on 15th when two were off the Farne Islands. Few were evident in August with only five N recorded at Seaton Sluice between 9th-27th and singles at Hauxley and Annstead Point during the same period.

Numbers continued to be low in September with three off Hauxley on 5th when two were also noted on a pelagic trip from Blyth. Singles also moved N at St Mary's Island on 23rd and at Newbiggin on 26th.

During October movement increased with the monthly total at Seaton Sluice being 33 N and two S. Peak passage occurred on 9th when 25 moved N off Cullercoats and 20 N off the Farne Islands. This passage continued on 10th with 11 N off Newbiggin. One-five were also noted from St Mary's Island, Cresswell, the Farne Islands and Holy Island up to the 25th.

**Manx Shearwater,** *Puffinus puffinus*
A common passage visitor.

Very light spring passage began in April with one-two off Seaton Sluice and the Farne Islands from 15th. Numbers increased in May with 53 N at Seaton Sluice during the month with lighter passage noted at seven other sea-watching sites.

In settled weather conditions numbers were lower than usual during the mid-summer period. During July the monthly total at Seaton Sluice was only 379 N and 33 S while during August only 138 were recorded flying N and 32 S. Similar numbers were also noted at Seaton Sluice during September.

Few were seen in October, the last being a single off Snab Point, Cresswell, on 14th November.

**Storm Petrel,** *Hydrobates pelagicus*
An uncommon passage visitor.

Once again the use of tape recordings by ringers confirmed the presence of the species offshore during the summer.

Most records came once again from the efforts of ringers operating at Tynemouth. A bird trapped there on 12th June was the earliest ever caught at the site. Thereafter 103 were caught between June-August (168 in 1991), despite a similar level of effort. One individual trapped in August had been ringed at Tynemouth in the same month of 1991.

Away from this locality an unusual record was provided by a single found fluttering around the cottage on Brownsman, Farne Islands, on 29th September.

Ringing once again demonstrated the nomadic nature of the species. For example, a record from 1991 involves an individual marked at Flamborough, N Yorkshire, on 19th July being controlled at Tynemouth on 2nd August, at Hauxley on 23rd August and then in Norway, on 25th August 1992. During 1992 there was an example of rapid movement between sites with an individual ringed at Hauxley on 8th August being controlled one hour 45 minutes later at Whitburn, Tyne and Wear. Finally, one trapped at Tynemouth during the summer had a Spanish ring. Full details are awaited from Spain where there have been recent surveys in the Balearic Islands.

**Leach's Petrel,** *Oceanodroma leuchorhoa*
A rare visitor.

One was trapped at Tynemouth on 9th August. The other records followed NE winds in October when singles flew S at St Mary's Island on 9th and N at Seaton Sluice and Newbiggin on 10th.

**Swinhoe's Storm-petrel,** *Oceanodroma monorhis*
An extremely rare visitor.

One of the undoubted highlights of the year, not just for the county but for Britain, was the identification of the dark-rumped petrel caught at Tynemouth in July 1990, again in July 1991 and, incredibly, once again on 30th July.

A special paper on its identification appears later in this report.

**Gannet,** *Sula bassana*
An abundant summer and passage visitor.

Early records involved one-four off Hauxley, Newton Point and Holy Island from 21st February. More normal spring passage followed in March with 230 per hour N at Seaton Sluice on 23rd and 190 per hour N at Newbiggin on 27th.

Movement continued during April while during May heavy passage N was again noted with 400 per hour on 2nd off Newton and at the Long Nanny Burn on 21st.

During the summer months feeding parties from the Bass Rock were regularly noted offshore.

Further heavy movement occurred during August with 500 per hour N at Hauxley on 22nd and an even more impressive 1,139 moving hourly at the Farne Islands on 29th. Similar strong passage continued in early September with 1,033 N per hour at Newton Point on 5th and 744 N at the Farne Islands on 23rd.

In October the heaviest movements N involved 735 at Cullercoats on 9th and 800 per hour at Annstead Point on 20th. Numbers fell rapidly in November while in December only one-three were seen from Newbiggin and Holy Island on a number of dates.

**Cormorant,** *Phalacrocorax carbo*
A common visitor and well-represented breeding species.

During the first quarter the peak counts were at roosting sites. 67 were at Blyth Harbour, 66 at North Shields Fish Quay and 59 at Amble. Inland there was a January peak of 65 at the regular Corbridge roost. Birds showing characteristics of the Continental race, *P. c. sinensis* were on passage in February-April with one-eight at seven coastal and four inland localities.

262 pairs bred at the Farne Islands, the same number as in 1991. The first eggs were seen on 29th April and most young had fledged by 11th July. 24 pairs bred on Big Harcar, the first time this island in the group had been used since 1981.

Post-breeding dispersal followed in August when 42 roosted at Castle Island. During the final quarter the favoured localities provided the largest counts with 66 at North Shields Fish Quay, 62 at Blyth and 58 at Amble harbour. The Corbridge roost held 40 in December.

**Shag,** *Phalacrocorax aristotelis*
A well-represented breeding species and a common visitor.

Most of the population was concentrated around the Farne Islands throughout the year and regular movements to feeding grounds off the N shore of Holy Island were regularly noted.

Passage was indicated by 51 moving N off Seaton Sluice during March before birds settled down to breed. The Farne Islands breeding colonies had another record year with 1,871 pairs (1,716 in 1991). The first young were seen on 4th May but, as usual, the season was protracted with the last fledging during the first week in October. The only other breeding records were six pairs at Dunstanburgh and three occupied nests at Needles Eye, near Berwick, in June.

Dispersal from the Farne Islands colonies occurred in September when 205 moved S off Seaton Sluice and 175 flew N. Smaller feeding parties were spread right along the coastline during autumn and winter.

An individual ringed on the Farne Islands in 1975 was found dead in Grampian in June, aged 17 years.

**Little Egret,** *Egretta garzetta*
An extremely rare visitor.

One spent five hours in the Cresswell Pond, Druridge Pools and E Chevington areas on 5th May. This was the seventh county record, the last being in Budle Bay and the Tweed Estuary in June-July 1990.

**Great White Egret,** *Egretta alba*
An extremely rare visitor.

One was at Holywell Pond on 4th July. This was the third county record, the last being an individual seen moving over Whitley Bay to Holywell Pond and then Druridge Pools in May 1989. The latest sighting is still under consideration by the British Birds Rarities Committee.

**Grey Heron,** *Ardea cinerea*
An uncommon breeding species, well-represented as a passage visitor.

The largest winter gathering between January-March involved 29 on the Tweed at Berwick with one-11 being noted in at least 14 other localities.

Dispersal to breeding areas occurred in March but the trend of recent years for pairs to nest either singly or in very small groups continued to make monitoring very difficult. Two occupied nests in the S of the county had young by 16th April while late breeding was indicated by a nest on the Otterburn Ranges with two young on 8th July.

Post-breeding dispersal led to a gathering of 20 at one of the favoured sites, Castle Island, by 27th June, increasing to a peak of 37 by late August.

During the final quarter the biggest concentration involving 40 birds was again at Berwick and one-nine were seen at a further 13 localities.

An example of longevity was provided by an individual found dying at Consett, Co. Durham, in October which had been ringed as a pullus at Slaley 21 years earlier.

**Spoonbill,** *Platalea leucorodia*
A rare visitor.

One visited Hauxley Reserve on 14th May and another remained in Budle Bay from 25th July-3rd August.

**Mute Swan,** *Cygnus olor*
An uncommon breeding species, well-represented as a non-breeder, passage and winter visitor.

Favoured areas attracted the biggest herds during the first quarter with peaks of 205 at Berwick, 78 at Wark on the Tweed, 38 at Killingworth Lake, 30 at Blyth Harbour and 18 at the Queen Elizabeth II Park, Ashington.

Pairs returned to breeding territories later than usual due to poor weather conditions. However, by April 102 pairs were located and of these, 55 pairs nested and 39 hatched at least 149 cygnets. The pairs for which a hatching success was known produced an average of 2.9 young compared with 4.2 in 1991, a reflection of the poor weather of March-April when many nests were flooded.

Numbers at the pre-moulting Ashington herd rose to 33 in June while the peak count of moulting birds at Berwick was 675 in late July.

In the latter part of the year numbers at the regular sites rose to 42 at Queen Elizabeth II Park, 38 at Killingworth Lake, 32 at Castle Island, 27 at Blyth and 21 at Amble. Other counts for this period included 21 at Wark, 20 at Shield-on-the-Wall and 15 at Lindisfarne. The Berwick herd decreased to 288 by December.

Ringing again provided much information about movements. In July 386 birds were caught at Berwick and 188 required ringing. These birds have since been sighted as far S as York, N to Edinburgh and W to the Clyde. Many others were recorded within the county and along the Tweed and Teviot.

In the county a further 58 non-breeders and 78 cygnets were ringed. The most notable movements involved a bird ringed at Amble in May 1991 moving to Berwick in July and then being seen at Wroxham, Norfolk, in July 1992. Another ringed at Ashington in May 1991 was at Welney, Norfolk, in January 1992.

A 'Polish morph' was ringed at Tynemouth in April, the first confirmed county record for this variation.

**Bewick's Swan,** *Cygnus columbianus*
An uncommon passage and winter visitor.

A single bird was in the SE during the first quarter. It was at Castle Island on 1st January and then joined Whooper Swans, *Cygnus cygnus*, in the Linton area from 8th January-20th March. This individual, or perhaps another, was in the Cresswell Pond-Druridge Pools area between 20th-25th April.

The species was then absent until return passage in late October when one visited Ellington on 29th and a family party of four was at Cresswell Pond on 31st. In November three adults were at Druridge Pools on 1st and one remained with Whooper Swans in the Cresswell Pond, Ellington and Warkworth Lanes area until at least 29th. It or another was at Holywell Pond on 30th November.

**Whooper Swan,** *Cygnus cygnus*
A well-represented passage and winter visitor.

The favoured wintering area around Druridge Bay produced a peak of 98 at Linton in February. Elsewhere during the first quarter there were herds of 82 at Lindisfarne, 45 at Cornhill and 37 near Wark on the Tweed and many records of small parties in other localities.

Numbers quickly declined in April with movement N to breeding areas while a single at Scremerston in June may have been injured or sick and unable to migrate.

Return movement began in October with a single at Grindon Lough on 8th, increasing to 30 by 31st. Other arrivals led to a peak of 84 in the Druridge Bay and Ashington areas by 8th November. The Grindon Lough herd increased to 38 during November while the county's largest late concentration was 100 at Cornhill in late December. Parties of one-11 were in at least a further ten localities.

Ringing once again provided data indicating the random nature of wintering parties. The Linton herd in December included five birds ringed on breeding grounds in Iceland between 1988-92 and one marked at Welney, Norfolk, in 1991. Four of the Icelandic birds had been identified previously in other areas. One had been in Denmark in 1990 while others were in Ireland, in Co. Tyrone in 1990-91 and Co. Donegal in the winter of 1991-92 while another had wintered in Cumbria in 1990. One of the birds ringed in Iceland in 1988 had also been seen at Cresswell in 1991-92.

Two adults at Grindon Lough in October had been ringed in Iceland in July 1991 and at Caerlaverock, Dumfries and Galloway, in November 1989 respectively. An adult ringed at Killingworth Lake in October was sighted at Wheldrake Ings, Yorkshire, in December.

**Bean Goose,** *Anser fabalis*
An uncommon passage and winter visitor.

One flying S at Cocklawburn on 26th September was the only sighting.

**Pink-footed Goose,** *Anser brachyrhynchus*
A well-represented passage visitor.

During the first quarter small parties were present with 35 at Grindon Lough, 32 in Budle Bay, 26 seen flying NW at Big Waters and 25 in the Rayburn Lake area.

Movement through the region led to an increase in April when the Rayburn Lake flock increased to 53. 23 visited Warkworth Lane Pond, 21 passed through Holy Island and 20 were at Holywell Pond. Parties of one-12 were noted at six further localities.

During May-July one-two non-breeders remained at ten widespread localities. During August five were at Holywell Pond and a single at Druridge Pools.

The first indication of return passage was on 31st August when 37 flew SW over Jesmond Dene. Light passage continued in September, building up to concentrations in October of 220 in Budle Bay and 100 on Holy Island. Continued immigration in November produced 50 over Warkworth Lane Pond on 4th, 200 SE over Thrunton on 7th and 420 at Lindisfarne on 15th.

Onward passage reduced numbers in December when the largest groups comprised 150 at Lindisfarne, 70 near Stannington and 20 at Caistron.

**White-fronted Goose,** *Anser albifrons*
An uncommon visitor

During January two were with Greylag Geese, *Anser anser*, in the Cresswell and Queen Elizabeth II Park area between 3rd-19th. Four found at Grindon Lough on 2nd increased to 16 by 26th February. This party, all of the Greenland race, *A. a. flavirostris*, remained until late April. The only other early reports involved singles at Holywell Pond between 19th-21st February and at Derwent from 8th-21st March.

The species was then absent until three adults returned to Grindon Lough on 24th October and were joined by two more by 28th November. The only other record was of three flying NW at Druridge Pools on 31st December.

**Greylag Goose,** *Anser anser*
A rare breeding species but a common passage and winter visitor.

Regular wintering haunts held large concentrations during January-April with 2,450 around Budle Bay, 750 near Wooler, 715 at Caistron, 650 at Hallington and parties of 380-520 regularly at Whittle Dene, Grindon Lough, Holywell Pond and Rayburn Lake.

Breeding occurred at three localities. At Caistron five broods comprised 23 young, at Holywell Pond there were three broods of 12 young and at Long Nanny Burn a pair raised three goslings.

Autumn arrival began in September with 44 E over Longbenton on 3rd. On 4th 122 gathered at Holywell Pond and by 18th 48 had returned to Caistron. The population continued to increase in October to provide final quarter peaks of 2,500 at Holburn Moss, 1,500 at Budle Bay, 680 at Caistron, 420 at Holywell Pond and Little Swinburne and 400 near Derwent.

**Canada Goose,** *Branta canadensis*
An uncommon breeding species, well-represented as a non-breeder and a passage and winter visitor.

The usual favoured reservoirs and large lakes held the main concentrations with parties frequently moving between feeding areas. Peak counts during the first quarter involved 157 at Hallington, 55 at Capheaton Lake and one-19 at a further 12 localities.

Breeding followed on several waters. Two pairs raised 17 young at Caistron and there were single broods at Castle Island and Coanwood (five young at each site), Gosforth Park (four young), Colt Crag (one young) while two goslings were noted at a moorland site in the SW.

Movement towards Scottish moulting areas followed with 120 moving N at Lindisfarne in late June while in July 200 were noted at Colt Crag. On 31st August 300 fed in stubble near Hallington, the largest concentration ever recorded in the county. This flock may have included birds on passage after the moult migration.

During the final quarter flocks once again moved between the favoured feeding areas with peaks of 265 at Colt Crag and 229 at Capheaton Lake.

The injured bird first recorded in 1987 at the Farne Islands remained throughout the year.

**Barnacle Goose,** *Branta leucopsis*
A well-represented passage and winter visitor.

During the first quarter 13 were regularly in the Budle Bay area, five were seen flying S at St Mary's Island and two were settled at Caistron.

Spring passage towards Arctic breeding ground, presumably of birds from the wintering flocks on the Solway, occurred in April with 200 moving E over Gilsland on 4th. During May-June single non-breeders remained at three coastal and three inland localities.

Return passage began in September with two at Druridge Pools on 6th and 20 moving W over Housesteads on 29th. Movement peaked on 3rd October with up to 1,000 in Budle Bay, 350 at Ladyburn Lake, 230 at Hauxley, 200 at St Mary's Island and Tynemouth and 40-192 at six other localities. Rapid passage through the area led to a quick decline in numbers when, except for 82 S at Seaton Sluice on 5th, only one-eight were seen at four other localities up to 25th.

A second period of lighter passage occurred in late November with 120 over Budle Bay on 22nd. In December 70 moved SW over Heaton on 9th and 16 flew N at Druridge Bay on 22nd. One-five were also seen during December at three other localities.

**Brent Goose,** *Branta bernicla*
A well-represented passage and winter visitor.

Numbers were low during the first quarter with a peak of only 200 birds of the pale-bellied race, *B. b. hrota*, at Lindisfarne in February. One-two birds of the dark-bellied race, *B. b. bernicla*, were seen during this period at St Mary's Island, Cresswell, Hauxley, Alnmouth and Bamburgh.

Most had departed by April when six dark-bellied birds were at Lindisfarne on 28th and two at Warkworth and Longhoughton Steel. An unusual mid-summer record was provided by a single dark-bellied bird offshore at Cheswick Black Rocks on 12th July.

None were seen in August but the now usual early arrival back at Lindisfarne began in early September. 85 were present on 6th, rapidly increasing to 900 by 30th. Further arrivals swelled the local population to 1,700 during October, gradually rising to a late peak of 1,850 in December. The increasing trend of dark-bellied birds to visit at Lindisfarne was again noticed with up to 350 present during November.

Elsewhere during the final quarter only one-five dark-bellied birds were seen at four coastal localities.

**Shelduck,** *Tadorna tadorna*
An uncommon breeding species, well-represented as a passage and winter visitor.

The rich feeding provided by Lindisfarne, in particular Budle Bay, attracted the main population during the first quarter with a peak of 1,330. Elsewhere, larger groups involved 49 in the Coquet estuary, 45 at Cresswell Pond-Druridge Pools, 39 at Blyth and 36 at Hauxley Reserve.

Breeding followed in typical localities and the local results from a survey organised by the Wildfowl and Wetlands Trust appear as a special paper later in this report.

Post-breeding gatherings occurred in September with 160 at Lindisfarne, 40 at Hauxley and 16 at Druridge Pools. Numbers increased in the final quarter to provide a December peak of 580 at Lindisfarne where once again most were in Budle Bay.

There were several inland records during the year away from two non-coastal localities where breeding occurred. They involved one-two visiting Big Waters, Derwent, Grindon Lough and Wark.

**SHELDUCK AND BROOD** – one of the species surveyed during the year

**Wigeon,** *Anas penelope*
An abundant passage and winter visitor and a rare breeder.

Larger concentrations during the first quarter included 750 in Budle Bay and 600 at Cresswell Pond while feeding areas inland held 535 at Derwent and 416 at Grindon Lough. Numbers quickly declined in early spring as birds moved away to breeding areas.

Despite one-five pairs being located in five suitable areas successful breeding occurred at only two localities. This compared with five sites during 1991 which had given hope of an expansion of our small breeding population. The areas involved were Derwent where there were 11 broods totalling 48 young while at another locality four pairs fledged only nine young.

Light return passage was noted in July-August. Heavier arrivals occurred in September when numbers at Lindisfarne increased from 350 on 6th to 9,000 on 30th. During the month parties moving towards Lindisfarne produced counts at Seaton Sluice of 455 N while only 38 moved S. Passage increased in October with a monthly total at Seaton Sluice of 1,433 N and 25 S, the maximum movement being on 10th when 632 passed N.

Numbers at Lindisfarne reached a late peak of 15,500 on 21st October. Elsewhere during the final quarter the maximum counts involved 382 at Grindon Lough, 270 at Cresswell Pond and 258 at Colt Crag.

**Gadwall,** *Anas strepera*
An uncommon visitor and a casual breeding species.

Up to 13 frequented Ladyburn Lake in January while elsewhere during the first quarter one-four were at eight other localities.

During April-May five pairs were at Caistron, four pairs at Grindon Lough, seven individuals were regularly at Druridge Pools and one-three occurred at seven other localities, including a pair on the Farne Islands on 13th May.

Successful breeding has been recorded on only three previous occasions, the last in 1988. However, this year at Caistron there were three broods comprising 24 young while at a site in the W a nest containing 11 eggs was found but the outcome was unknown.

Peak counts for the September-December period involved six at Ladyburn Lake, five at Caistron and four at Hauxley Reserve. One-three occurred at a further nine localities.

**Teal,** *Anas crecca*
An uncommon breeding species but a common passage and winter visitor.

Peak counts during the first quarter involved 306 at Grindon Lough, 294 at Holywell Pond, 261 at Whittle Dene and 186 in Budle Bay.

During March pairs were back in nesting areas, including high moorland. Breeding followed and broods were reported from Caistron (three), Derwent (two) with single broods at Arcot Pond, Holburn Moss and Ottercops.

Post-breeding gatherings and the early arrival of passage birds led to peaks in September of 260 at Caistron, 210 at Big Waters and 180 at Holburn Moss. Typical offshore movements were also seen with a September total of 565 N and 306 S at Seaton Sluice.

During the final quarter the largest gatherings involved 496 at Lindisfarne, 480 at Holburn Moss, 408 at Caistron, 310 at Wallsend Swallow Pond and 282 at Big Waters.

A drake ringed at Blaydon in November 1987 was killed by a Mink, *Lutreola lutreola vison*, at Kielder in April 1991.

Birds of the N American race, Green-winged Teal, *A. c. carolinensis*, are extremely rare visitors. However, a drake showing characteristics of this race was at Holy Island Lough on 1st February. Another frequented Wallsend Swallow Pond between 24th October-1st December and was also at Holywell Pond on 14th November. These provided the seventh and eighth county records.

**Mallard,** *Anas platyrhynchos*
A well-represented breeding species and an abundant passage and winter visitor.

Favoured wintering and feeding areas held the main concentrations during the first quarter. 1,065 were at Lindisfarne in January while elsewhere 373 were at Whittle Dene, 354 at Big Waters and 310 at Holywell Pond.

Many pairs were back in regular breeding areas in March and the first brood of young was noted at Wark by 25th April. Successful breeding was recorded from at least 22 localities. Among the more regularly monitored sites were Holywell Pond (six broods, 63 young), Caistron (nine broods, 47 young), and Derwent (five broods, 49 young).

Post-breeding gatherings and influxes of passage birds in August-September involved peaks of 670 at Caistron, 470 at Whittle Dene, 411 at Big Waters and 405 at Derwent. Further slight increases occurred during the final quarter to produce peak counts of 510 at Big Waters, 453 at Derwent, 420 at Lindisfarne and 300-350 at Colt Crag, Whittle Dene and Caistron.

**Pintail,** *Anas acuta*
A well-represented passage and winter visitor.

Numbers at the favoured wintering locality, Lindisfarne, peaked at 41 on 19th January. Elsewhere during the first quarter 12 were at Grindon Lough, five at Holywell Pond and one-three at another eight localities.

Numbers quickly decreased in April-early May with movement back to breeding regions. The only summer record involved a drake at Big Waters between 13th-21st June.

A party of 20 flying N at Hauxley on 22nd August heralded a return and by September 11-12 had gathered at Hauxley Reserve and Druridge Pools, ten were at Derwent and seven at Whittle Dene.

During the final quarter peak counts involved eight at Derwent, five at Lindisfarne and three at Holywell Pond. One-two were seen at a further six localities.

**Garganey,** *Anas querquedula*
An uncommon passage visitor and a rare breeding species.

The first spring arrivals occurred in April with two at Wallsend Swallow Pond on 6th, a pair at Big Waters on 23rd and a drake at Druridge Pools on 27th.

Numbers increased in May with pairs at Big Waters, Whittle Dene, Greenlee Lough, Druridge Pools, Coquet Island and Holy Island Lough. One-three drakes also occurred at a further six localities, including the Farne Islands.

Pairs remained at Big Waters and Holy Island Lough in June with drakes also noted at Marden Quarry, Druridge Pools and Newton Pond. In July females or immatures occurred at Druridge Pools and Hauxley Reserve.

The final records were in August and involved two at Bell's Pond, Cresswell, on 8th and singles at Whittle Dene on 9th and Druridge Pools on 24th.

**Shoveler,** *Anas clypeata*
An uncommon breeding species and winter visitor, well-represented as a passage visitor.

Gatherings of 20 at Big Waters and 16 at Wallsend Swallow Pond in January were the largest parties during the first quarter when one-ten were seen at nine other waters.

Passage led to an increase in April when the ideal feeding provided by floodwater at Druridge Pools attracted 26. Five pairs were at Holy Island Lough where display took place. Four nests were later noted at the latter site but three were predated and only a single brood was raised. Results were better at Grindon Lough and Caistron with two broods fledging at each site

Light influxes of wintering birds led to peaks of 52 at Big Waters and 20 at Grindon Lough by October, while 30 were at Wallsend Swallow Pond in November. Parties of one-16 were noted in several other localities until the end of the year.

**Pochard,** *Aythya ferina*
A rare breeding species, common as a passage and winter visitor.

Numbers were generally low between January-April with peaks of only 59 at Druridge Pools, 50 at Ladyburn Lake, 48 at Whittle Dene and 44 at Hallington.

The departure of wintering birds led to further declines by May before breeding took place at four localities. These were Holywell Pond (five broods, 26 young), Arcot Pond (two broods, 12 young), Wallsend Swallow Pond (two broods) and a single brood of six at Annitsford Pond.

Return passage quickly swelled numbers from August, leading during September to peaks of 99 at Whittle Dene and 84 at Big Waters. During the final quarter gatherings of 83 were settled at Whittle Dene, 55 at Ladyburn Lake, 53 at Killingworth Lake and 30 at Big Waters.

**Tufted Duck,** *Aythya fuligula*
An uncommon breeding species, common as a passage and winter visitor.

Between January-April the largest wintering flocks involved 72 at Ladyburn Lake, 66 at Holywell Pond, 60 at Killingworth Lake and 50-55 at Hauxley Reserve, Whittle Dene, Hartburn and Cresswell Pond.

Breeding followed at many waters. Among the more regularly monitored sites there were nine broods at Caistron, seven at Holywell Pond, six at Wallsend Swallow Pond, five at Arcot Pond, three at Castle Island and two at Annitsford Pond. Single broods were also noted at other localities including Whittle Dene, Ladyburn Lake, Hauxley Reserve, Spindlestone and Holburn Moss.

Post-breeding gatherings, swelled by arriving passage groups, resulted in August gatherings of 194 at Whittle Dene and 72 at Colt Crag. During

September 200 were at Hauxley Reserve and 184 at Ladyburn Lake.

Onward passage led to a decline in the final quarter when the largest wintering groups involved 72 at Ladyburn Lake, 71 at Druridge Pools, 60 at Hauxley and 45 at Holy Island Lough.

**Scaup,** *Aythya marila*
A well-represented passage and winter visitor.

A party of ten at Hauxley Reserve was the largest group during January-April when one-three were seen at another four coastal localities. More unusual was the drake which appeared at Whittle Dene in October 1991 and remained until at least 2nd May, being joined briefly by another male on 11th April.

During May only four were seen flying N at Seaton Sluice and two S at Holy Island and a drake which appeared at Wallsend Swallow Pond on 28th remained until 7th June. In June single females also visited Big Waters and Whittle Dene, the latter remaining throughout July. Also in July one-three were seen at Hauxley, Ladyburn Lake and Seaton Sluice.

During August two females stayed at Big Waters from 2nd-27th, four flew N at Seaton Sluice on 14th and one-two were also reported from Whittle Dene, Druridge Pools and Hauxley.

During September the largest group involved only six at Alnmouth with one-three noted at six other localities. They were particularly scarce during the final quarter with singles being seen at only five coastal localities.

**Eider,** *Somateria mollissima*
A common breeding resident.

The largest counts during the first quarter were in January when 1,738 were between Budle Point-Seahouses, 1,030 were at Lindisfarne and 100-380 were at a further nine localities.

During April-May the leucistic individual found in August 1989 was again noted at Holy Island.

On the main breeding site, the Farne Islands, the decline of recent years was halted and 1,202 nesting females were counted (1,074 in 1991). The first eggs were noted on 2nd May and the first young on 31st May. On Coquet Island 364 nests were located (386 in 1991) with the first eggs on 16th April.

Post-breeding gatherings produced 2,380 at Lindisfarne in July. As usual most flocks remained in the N and during the final quarter the peak counts involved 2,038 at Lindisfarne, 1,300 around the Farne Islands and 1,280 between Budle Point-Seahouses. Further S the largest counts involved 450 off Cambois and 550 at Newbiggin.

While Eiders are essentially sedentary, one found on the Farne Islands during the year had been ringed in April 1981 on the Baltic coast of Germany. There have been previous examples of interchange between Northumberland, Denmark and Germany. A Farnes-ringed bird was also found dead on Tayside during 1992.

**King Eider,** *Somateria spectabilis*
A rare visitor.

A drake at Cheswick Black Rocks, Cocklawburn, on 27th-28th June was only the third county record this century, the last being a female found dead at Fenham Flats in January 1974. The record is still under consideration by the British Birds Rarities Committee.

**Long-tailed Duck,** *Clangula hyemalis*
A well-represented passage and winter visitor.

The favoured wintering area between Bamburgh-Lindisfarne held the main concentrations during the first quarter with peak counts of 70 off Ross on 25th January and 180 at Stag Rocks on 8th March. Elsewhere only one-ten were noted in five localities and at Hauxley Reserve and Ladyburn Lake.

During April 141 remained off Bamburgh on 11th before passage back to Arctic breeding grounds led to a quick decline. One-two remained to summer and were frequently seen at Hauxley Reserve.

Return passage began on 16th September when one flew S at Holy Island while from early October more regular sightings occurred, producing a monthly total of 31 N at Seaton Sluice. A party of 20 were off the N Shore, Holy Island, on 21st with smaller groups seen at eight other localities. Numbers peaked in November when 100 were off Ross.

An unusual inland record was provided by a single at Grindon Lough on 24th October.

**Surf Scoter,** *Melanitta perspicillata*
A rare visitor.

There were two records during the year. A drake was at Cheswick Black Rocks, Cocklawburn, between 21st-30th June and another was in Druridge Bay on 16th October. They brought the county total to 13 records, the last being in November 1990 off Seaton Sluice.

**Common Scoter,** *Melanitta nigra*
A common visitor.

During the first quarter 800 were regularly off Bamburgh while sightings of 200-350 noted off Newbiggin, Cambois and in Druridge Bay probably involved the same birds moving between feeding areas.

Numbers remained high in April-May when there was a peak count of 420 off Bamburgh, 300 were at Cocklawburn and 100 flew N at Long Nanny Burn.

Summer passage occurred in June at Seaton Sluice with 476 N and 82 S during the month while 450 remained at Cocklawburn. Movements increased in July when the monthly total at Seaton Sluice was 533 N and 90S. A further rise occurred in August with 786 N and 24 S at Seaton Sluice and then peaked in September with 1,132 N and 32 S. The largest gathering was 120 in Druridge Bay.

During the final quarter 400 were again wintering in the favoured area off Bamburgh and Ross.

The year's only freshwater record involved five at Bakethin on 25th March.

**Velvet Scoter,** *Melanitta fusca*
A well-represented visitor.

Small parties involving one-six were regular between January-April off Seaton Sluice, Blyth, Newbiggin, Druridge Bay and Bamburgh. Some remained in summer and between May-August, one-two being seen irregularly from five localities.

A slight increase was evident in September when nine were in Druridge Bay on 20th and one-two were also seen at a further five localities.

Conspicuous passage occurred between 4th-18th October. The October total at Seaton Sluice involved 100 passing N with lesser numbers reported from Cullercoats, Newbiggin, Hauxley, the Farne Islands and Holy Island. Gatherings of up to 50 occurred in October in Druridge Bay with 20 off Cambois. This was followed in November with a gathering of 152 off Cocklawburn, the biggest gathering ever recorded in the county. During December one-seven were again wintering in at least four localities.

**Goldeneye,** *Bucephala clangula*
A well-represented passage and winter visitor.

The usual favoured haunts on major rivers, reservoirs, lakes and estuaries attracted the main gatherings between January-April. Peak counts involved 385 at Berwick, 130 on the Tyne between Newburn-Corbridge, 90 offshore at Cambois, 80 at Capheaton Lake and 57 at Whittle Dene.

Movement back to breeding areas led to sharp decline in late April although, as usual, small numbers remained in summer.

Return passage was prominent in October with 503 N and 114 S at Seaton Sluice during the month. 100 flew N at Newbiggin on 10th during gales which prompted heavy sea passage of several species. During October inland gatherings formed with 100 at Hallington and 30 at Whittle Dene.

During November-December wintering flocks were again concentrated in good feeding areas with 264 at Berwick, 150 in Alnmouth Bay, 122 at Cambois and 107 on the Tyne between Newburn-Close House.

**Smew,** *Mergus albellus*
An uncommon winter visitor.

During January a drake was at Big Waters on 21st and one-two 'redheads' were also seen at this locality and at Killingworth Lake, Druridge Pools, Ladyburn Lake, Hauxley Reserve and on the sea off Whitley Bay. In February-March single 'redheads' continued to frequent these waters while another occurred at Tyne Green, Hexham.

An unseasonal individual was at Druridge Pools and Hauxley Reserve between 20th-26th May.

The species was then absent until October when a 'redhead' flew N at St Mary's Island on 15th and another appeared at Arcot Pond on 24th.

During November sightings of 'redheads' at Big Waters, Killingworth Lake and Holywell Pond may have involved the same individual. In December 'redheads' at Killingworth Lake throughout the month and at Seghill reserve on 6th and 18th were considered to be separate individuals. Other singles occurred in December at Ladyburn Lake between 16th-31st and at Cambois on 30th.

**Red-breasted Merganser,** *Mergus serrator*
A well-represented passage visitor which has bred.

Off-shore waters held the main parties in the first quarter with 20-25 regularly in Druridge Bay while 19 were seen off Warkworth and 11 between Bamburgh-Budle Bay. Most of the Druridge Bay birds remained into April but numbers then declined sharply between May-July.

The main summer gathering involved 50 off Beadnell by 9th August and a similar number in Druridge Bay on 30th September although few were seen elsewhere.

Passage was evident in October when the monthly total at Seaton Sluice was 84 N and two S while parties of 35 were in Druridge Bay with some regularly visiting Ladyburn Lake, 34 off Holy Island and nine at Cocklawburn.

During November-December the largest groups involved 42 frequenting Druridge Bay, 20 in Alnmouth Bay, 15 between Newbiggin-Cambois and 14 at Ladyburn Lake.

Inland records involved one on the Coquet at Sharperton in March, three at Colt Crag in September and a single at Bakethin in November.

**Goosander,** *Mergus merganser*
A well-represented breeding species and a passage and winter visitor.

The largest gatherings during the first quarter were 90 at Wylam, 49 at Whittle Dene and 31 on the Tweed at Cornhill. Up to 38 remained on the Tyne at Close House until 3rd April.

Breeding then occurred and in May broods were noted at Mitford, Cupola Bridge, Riding Mill, Tyne Green and Bywell Bridge. Further success was recorded in June from Ovingham, Kielder and Hexham. Three broods totalling 30 young were found at Derwent.

Small groups gathered inland during the summer and autumn, the largest party being 23 at Colt Crag on 22nd August. During the final quarter the usual favoured feeding areas attracted the main concentrations with up to 45 at Whittle Dene, 31 at Wylam and 24 at Caistron.

**Ruddy Duck,** *Oxyura jamicensis*
A rare visitor which bred for the first time in 1991.

A single at Ladyburn Lake in January was the only early record but by the end of March a pair was at Holywell Pond and a female at Capheaton Lake.

An influx occurred in April when three pairs were at Holywell Pond, two pairs at Big Waters and one-two at a further five localities. During May they were seen at 14 waters before successful breeding occurred at five localities. This involved three broods totally 19 young at Holywell Pond, two broods comprising nine young at Annitsford Pond and single broods at Arcot Pond and Wallsend Swallow Pond and Hartburn Lake.

Post-breeding gatherings involved peaks of 15 at Hauxley Reserve in July, 13 at Holywell Pond in August and one-five at a further ten localities.

During the final quarter one-three were seen at Druridge Pools, Arcot Pond, Hartburn Lake, Holywell Pond, Colt Crag, Capheaton Lake and on the Tweed at Berwick.

**Honey Buzzard,** *Pernis apivorus*
A rare visitor.

One flew NW over Benton on 9th May and another drifted SW over Holywell on 4th July, providing the first records since May 1988.

**Marsh Harrier,** *Circus aeruginosus*
A rare passage visitor.

Spring passage began in April with adult males over Harwood on 24th and Ellington on 25th. On 26th a tired immature female rested on rocks on the Inner Farne for ten minutes before flying towards the mainland.

During May a first-year male circled the Farne Islands on 13th, a young male was at Long Nany Burn on 14th and females were at Hadston and Long Nanny on 15th. Males followed at Holywell and Warkworth Lane ponds and Hauxley on 17th while further females passed through Holywell on 25th and

Long Nanny on 26th. The only June record involved a female over Wark Forest on 4th.

The only autumn sightings involved a female or juvenile which frequented Newton Pool from 1st-3rd October.

**Hen Harrier,** *Circus cyaneus*
A rare breeding species and an uncommon passage and winter visitor.

Moorland, forest edges and coastal dunes attracted wintering individuals during the first quarter. Males were noted in the Tarset, Grindon Lough and Wooler areas while 'ringtails' frequented Ross and several typical upland inland localities.

Spring passage produced single 'ringtails' at Holywell and Holy Island in April-May. Three pairs remained to breed in the county, fledging broods of six, three and one respectively. A pair failed at another site while in another locality a pair was driven off by a gamekeeper repeatedly crossing the site firing a gun.

An unusual record well away from any known breeding site was provided by an adult male hunting over forest in the Cheviots in late June.

Return passage commenced with a single 'ringtail' being heavily mobbed by crows, *corvid sp.*, near Craster on 30th September. During October at least one adult male was settled in the regular wintering area around Greenlee and Grindon loughs and Wark Forest.

During November there were sightings of an adult male at North Charlton and 'ringtails' were seen at Slaley and Druridge Bay. In December a male was noted on two dates in the Druridge Bay area and others were seen at Heighley Gate, near Morpeth, and Horton Grange while 'ringtails' were in two forest localities.

The Scottish origins of our passage and wintering birds was again demonstrated. An immature male at Harwood in April had been wing-tagged at a nest in Argyll in 1991. Females wing-tagged in nests in SW Scotland in 1991 and 1992 were also seen in the S Tyne Valley in November-December.

**Montagu's Harrier,** *Circus pygargus*
A rare visitor, and occasional breeder.

A female was seen in the W in late May. Following this sighting there were no more records until early August when a female with three newly-fledged young was watched in the W, provided the first confirmation of breeding in Northumberland since 1966.

**Goshawk,** *Accipiter gentilis*
A rare breeding species.

Birds were noted displaying and hunting in several forest and moorland areas during the year.

More unusual was a single which spent January-February in a lowland conifer plantation near Ponteland where kills included a Kestrel, *Falco tinnunculus*, Lapwing, *Vanellus vanellus*, and Woodpigeon, *Columba palumbus*.

Breeding was again recorded but, as in previous years, details are being withheld at the request of observers because of the continued threat to this very sensitive species from illegal persecution and thefts of eggs and young.

During the final quarter pairs and singles were reported from at least seven areas.

**Sparrowhawk,** *Accipiter nisus*
A well-represented breeding species and a passage and winter visitor.

The increasing tendency of wintering birds to carry out ambushes at garden feeding stations was recorded from many areas during the first quarter. Two regularly hunting a garden at Walbottle left it almost devoid of small birds by the end of January while in March a particularly bold female killed a Blackbird, *Turdus merula*, within feet of a gardener in Newcastle.

During the quarter the species was also well reported from a wide range of other habitats before moving into breeding territories during late March-early April. A female on the Farne Islands on 5th April suggested emigration. Breeding was recorded from many areas and in the South Central and Coquetdale study area four monitored nests fledged 16 young. In the SW one nest held six young, only the third occasion that a brood of this size has occurred in that area. Pairs were also recorded breeding in at least ten other localities. Dispersal from breeding areas occurred in late July-August when birds were noted back hunting in gardens in three areas.

The only suggestion of immigration was provided by a tired bird arriving from E at Seaton Sluice in late October, although during this period singles were regularly noted on the Farne Islands. Two were resident on Holy Island in October, regularly hunting a roost of Starlings, *Sturnus vulgaris*, at the Lough.

During November-December several hunted similar huge roosts in Newcastle city centre and the species was again regularly reported from gardens as well as rural areas.

A pullus ringed at Kielder in June 1991 died hitting a window at Clitheroe, Lancashire, in January, a movement of 152 kms. Another pullus ringed at Canonbie, Dumfries and Galloway, in July was controlled at Hauxley in September.

**Buzzard,** *Buteo buteo*
A rare breeding species and an uncommon visitor.

A significant increase in sightings and pairs in breeding territories was one of the most encouraging features of the year.

During the first quarter wintering individuals were noted from at least ten widespread localities while in March pairs were seen in several suitable breeding habitats.

At least six pairs went on to nest compared with only two in 1991 and one in 1990. Three pairs fledged five young while two nests failed during incubation from natural causes and the other was believed to have been illegally destroyed by a gamekeeper. This was by far the most successful breeding season recorded and, just as encouraging, was the fact that other birds held a total of 14 potential breeding sites for varying periods, giving hope of continued expansion. Wandering single birds were also noted during the breeding season and remainder of the year from at least 16 localities. The results show that the species could become firmly established in Northumberland if illegal persecution could be overcome.

Random dispersal from breeding and other summering areas was apparent in September-October with one-three birds being noted in several localities. A group of three was noted in the S in December.

BUZZARD – an encouraging breeding expansion occurred

**Rough-legged Buzzard,** *Buteo lagopus*
A rare passage and winter visitor.
   A single at Thrunton Woods on 6th November was the first county record since October 1988.

**Golden Eagle,** *Aquila chrysaetos*
A rare visitor.
   There were two records: a juvenile was noted soaring over the Irthing Valley on 15th March while on 31st October an adult pair was seen in the N.

**Osprey,** *Pandion haliaetus*
A rare visitor.
   Spring passage began in April with a typical record on 13th of a bird attracted to Greenlee Lough where it remained for a week. Two were over Wark Forest on 16th.
   Sightings increased in May with singles noted at Cresswell on 8th, Wark on 9th, Big Waters, Wideopen and Long Nanny Burn on 14th, again at Cresswell on 16th and Staward Gorge on 17th. Sightings at Bakethin on 20th and 22nd may have involved the same individual. On both occasions it was hunting and was seen to take two fish. None were seen in June but an adult in the S Tyne valley on 7th July may have been summering in the area.
   Return passage occurred in September with singles flying S offshore at Druridge on 13th and over Wark on 28th.

**Kestrel,** *Falco tinnunculus*
A well-represented breeding species and a passage visitor.
   During the first quarter one-two were reported from many localities ranging from Newcastle city centre to remote moorlands and hill districts. Six were along a two-mile stretch of the Roman Wall near Housesteads in January.

Display was noted at Harwood as early as 9th February with pairs being increasingly prominent at regular breeding sites during March-early April. During April singles occurred on the Farne Islands on four dates suggesting emigration.

During May singles and pairs were recorded in 33 localities. Breeding followed in many areas with the first fledged young noted by 19th June. The NW upland study area was less intensively covered than in recent years but four monitored sites fledged young. A fifth pair, using an old burrow of a Rabbit, *Oryctolagus cuniculus*, on a heather-covered bluff, failed during incubation, probably when one of the birds was predated.

Breeding also occurred in more urban areas with a pair fledging four young from a church site in the SE, used for the past 18 years. Other successful pairs used a dockside installation at Blyth and a coastal quarry.

During August-December they were again commonly distributed in virtually every suitable hunting habitat throughout the region with one-two regularly noted on the Farne Islands. Mild weather in October prompted a pair into prolonged and noisy display on the tower of St Nicholas Cathedral, Newcastle.

A pullus ringed at Elsdon in May 1989 was killed when it hit wires at Sunderland in January.

**Merlin,** *Falco columbarius*
An uncommon breeding species and a passage and winter visitor.

Coastal wintering haunts provided most records during the first quarter. There were many reports of individuals hunting a wide range of passerines, the most unusual being a wintering Chiffchaff, *Phylloscopus collybita*, which had a narrow escape at Newton Pool.

A return to inland areas was evident in February with singles, including four adult males, at seven localities before a more general movement back to breeding areas during March. During April only two were noted on the coast as pairs settled down to breed although a female remained on the Farne Islands from 5th April-8th May.

Monitoring by the Northumbria Ringing Group showed a total of 59 occupied sites early in the season but there were the usual failures because of poor weather or other natural causes. 36 pairs went on to fledge about 118 young, making it an average year for the species. Human interference continued to be a problem with a clutch taken at one site, where a notorious egg-collector was seen, and a brood of small young stolen on another moor.

The usual quick dispersal from breeding areas was indicated in August when both adults and juveniles were noted in four coastal areas. Others followed in September-October when they were distributed in good hunting areas with two-three at Lindisfarne and singles in at least nine other coastal areas.

During October-November singles were noted on many dates on the Farne Islands where prey included Turnstone, *Arenaria interpres*. During November-December they were well-reported from coastal haunts but with individuals also frequenting more urban areas with sightings, for example, from W Denton and Fenham in Newcastle.

A bird ringed as a pullus in the W in July was killed hitting wires at Kirtlington, Oxfordshire, in October, a movement of 357 kms. Another youngster, also ringed in July, was found dead at Dunbar, Lothian and Borders, in late August, 128 kms from the nest site.

**Hobby,** *Falco subbuteo*
A rare passage visitor which bred on one occasion.
 Spring passage in May produced singles at Threestoneburn on 11th and Howick Dene on 25th.
 No more were seen until August when there were three sightings. These involved individuals hunting over the walled garden at High Hauxley on 12th with perhaps the same bird being noted at Low Hauxley on 23rd. The other sighting was at Holywell on 16th.

**Peregrine,** *Falco peregrinus*
A rare breeding species and an uncommon passage and winter visitor.
 There were regular sightings from favoured coastal wintering areas, particularly Lindisfarne and Druridge Bay, throughout the first quarter when a male and female were semi-resident on the Farne Islands. However, others remained inland. More unusual records involved individuals mobbing a cock Hen Harrier, *Circus cyaneus*, at Grindon Lough, chasing a Cormorant, *Phalacrocorax carbo*, at Kyloe, and feeding on a Brent Goose, *Branta bernicla,* at Lindisfarne. It was not known whether it had killed the goose.
 Pairs were displaying around suitable breeding sites by February-March and during the season 16 sites were occupied. Seven pairs went on to fledge at least 17 young. One pair hatched five young, an unusually large brood, with four fledging. The fifth was found dead under the crag, perhaps after being accidently pulled from the ledge by an adult. One site was held by a pair which failed to lay eggs. Thefts of eggs and young continued to be a very serious problem and the remaining eight sites were all robbed.
 The usual quick dispersal was evident with a juvenile female at Cresswell by 6th August while between late-August-December there were many reports, mainly from coastal areas. At least two were settled for the winter at Lindisfarne, singles were noted on many dates on the Farne Islands, a juvenile was attracted to a wader roost at Newbiggin and another regularly roosted on the opencast mining spoil heap at Chevington. As usual, others remained inland with records from localities including Colt Crag, Grindon and Caistron where wildfowl and waders were available as prey while other winter singles were seen at Hulne Park and Kirkley Hall.

**Red Grouse,** *Lagopus lagopus*
A common moorland resident.
 Displaying pairs were holding territory on the Simonsides during January-March when most other reports were from the prime heather moorlands of Allendale and the Plenmeller area.
 During April-July they were recorded breeding in these and other moorland areas with many broods being noted, particularly on the Simonsides.
 During the final quarter they were distributed in typical habitats with parties being found in snow on the summit of Cheviot in late November and many pairs were displaying despite sub-zero conditions on the Simonsides in December. The largest late concentrations were in December with three flocks totalling over 60 at Coanwood, 35 at Williamston and 25 at Allenheads.

**Black Grouse,** *Tetrao tetrix*
A well-represented resident breeder.
 The main strongholds in the W provided virtually all records with the

first 'lek' of the year involving 26 males near Allenheads during March. A group of nine Greyhens was noted at Ninebanks during the same period.

The Allenheads 'lek' increased to hold a maximum of 32 cocks during April while three smaller gatherings on Wellhope Moor held 19 birds.

Sightings quickly dwindled as birds settled to breed. Most reports were again from the W with a scattering of individuals and pairs also being seen on the Otterburn ranges.

Post-breeding gatherings included up to 15 cocks and two-three Greyhens at Allenheads during September-October but only one-three were reported from other localities. Ten males remained near Allenheads in December.

**Red-legged partridge,** *Alectoris rufa*
An uncommon breeding resident introduced for shooting.

One-two were noted in the Hartburn and Netherwitton areas during February-April while a party of ten was at Swinhoe in May.

Breeding success was evident in August with an adult and two young at Caistron and a juvenile feeding on weed seeds in a garden at Grindonrigg in the Cheviots.

During the final quarter three were noted at Wallington in October, one-two at Hepple and Hartburn in November while 12 fed on weed seeds on gravel banks near Brandon in December.

**Grey Partridge,** *Perdix perdix*
A common resident.

Typical lowland arable fields and grassland provided most records with the largest early coveys involving 27 at Holywell and Newcastle Town Moor during January. An unusual sight was provided in February by two flying low over the Tyne from North Shields to South Shields.

During March pairs were noted in typical breeding areas at Horton and Arcot while during April-May they were reported from 29 suitable nesting areas with the first broods being noted on 26th June.

Post-breeding coveys occurred from August with numbers continuing to increase into autumn and winter. During the final quarter the largest coveys included 34 at Newbiggin, 32 at Woodhorn, 25 at West Chevington and Elsdon, with 15-18 noted from Harwood, Roseden, Stamfordham, Derwent and Prestwick Carr. The species' ability to survive in an urban area was demonstrated by a group of four noted regularly feeding on a grass verge in Scotswood Road, Newcastle, oblivious of heavy traffic.

**Quail,** *Coturnix coturnix*
A rare summer visitor and an occasional breeder.

Spring arrival was marked in May by a single flushed on the Inner Farne on 21st while on 26th another was in the Cheviots. During the first half of June ten birds calling at seven localities gave hope of a good breeding season.

During July individuals were heard in suitable breeding habitats among crops and grasslands at Netherwitton, Stamfordham, Howick, Long Nanny Burn and Lindisfarne while at Hetton, near Belford, two males called in competition. However, no young were reported to confirm successful nesting.

The only late record involved a pair flushed from a barley crop being harvested at Beanley on 11th September.

**Pheasant,** *Phasianus colchicus*
A widespread common resident reared and released in large numbers for shooting.

A concentration of 15 was around Wallsend Swallow Pond in January while during the first quarter individuals regularly visited feeding stations at Big Waters and Horton and a garden at Walbottle.

During the breeding season the first young were noted near Warkworth Lane Pond by 23rd April with success being reported from many other areas during May-July.

Suitable farmland feeding areas attracted parties during September-October. The largest late gatherings involved 57 near Derwent in late November and 50 in winter wheat near Cornhill in late December.

**Water Rail,** *Rallus aquaticus*
An uncommon passage and winter visitor and a rare breeding species.

Regularly watched localities including Big Waters, Gosforth Park, Holywell Pond, Caistron and Newton Pool all held one-two during the first quarter. Singles were also found at Barrasford and Linton and one crashed into a window at Seaton Sluice in February.

There were fewer records during the breeding season with singles at Bolam Lake, Newton Pool and Hauxley. Successful breeding is always difficult to prove but was suggested in August by juveniles at Ellington Pond and Big Waters while another at Whittle Dene was the first noted there for many years.

Immigration was indicated in September-October by a single which crashed into a window on Whitley Bay seafront and another was seen at Holy Island Lough.

During the final quarter one-two were again settled in regular winter haunts at Holywell Pond, Gosforth Park, Big Waters, Linton and Warkworth Lane ponds, Newton Pool and a single was on the Pont at Medburn.

**Corncrake,** *Crex crex*
A rare summer and passage visitor which formerly bred.

The only spring record involved an individual calling at Dunstan Stead, near Craster, on 2nd June.

On 5th August another was twice flushed during harvesting of oil seed rape near Belsay. The last record from this farm was in 1981 and the species had bred in the same field in the early 1960s.

**Moorhen,** *Gallinula chloropus*
A common breeding species.

Larger waters attracted the main concentrations during the first quarter with peaks of 52 at Holywell Pond, 22 at Caistron and Hauxley Reserve, 20 at Cresswell Pond and smaller groups at several other localities.

More unusual records in April involved singles at North Shields Fish Quay while another spent 5th-10th on the pond on Brownsman, Farne Islands.

Breeding followed in many suitable waterside habitats with the first broods noted from 13th May. Six broods at Caistron in July was the largest number reported. As usual, the season was protracted with small young at Grindon Lough in August and a half-grown chick at Howick as late at 19th October.

During the final quarter the main groups were in the favoured wintering localities with, for example, 22 at Holywell Pond, 15 at Caistron and smaller groups in at least six other areas.

**Coot,** *Fulica atra*
A well-represented breeding species and a passage and winter visitor.

Wintering concentrations were smaller than usual with peaks during the first quarter of 195 at Ladyburn Lake, 183 at Cresswell Pond, 72 at Caistron and 25-50 at seven other localities.

Numbers fell rapidly in late March-early April with the departure of wintering birds as local pairs settled into breeding territories. The first young were reported from Gosforth Park on 1st May while broods were also on many other waters in June-July.

Post-breeding gatherings began in August with 151 at Druridge Pool by 8th and with counts of 85 at Ladyburn Lake and 70 at Caistron.

Numbers continued to rise steadily with the arrival of wintering groups to produce peaks during the final quarter of 325 at Capheaton Lake, 250 at Cresswell Pond and Ladyburn Lake, 88 at Caistron, 77 at Hartburn Lake and up to 22 at five other waters.

**Oystercatcher,** *Haematopus ostralegus*
A well-represented breeding species and a common passage and winter visitor.

Suitable coastal feeding areas attracted the usual large early concentrations with a peak of 1,012 on the February wader count. Inland singles were at Derwent, Caistron and Reaveley in January and during February movement back to breeding areas was evident with 65 at Caistron by 16th when pairs were also at Grindon Lough and in four areas along the Tyne.

Coastal figures fell rapidly in March as the movement inland gathered momentum with pairs on territory in many areas and pre-breeding peaks of 205 at Caistron and 80 at Powburn.

Breeding followed in many suitable areas inland while on the Farne Islands 27 pairs bred, compared with 30 in 1991. The first eggs there were noted on 13th May and the first young on 23rd June. 23 pairs nested on Coquet Island compared with 17 in 1991 with first eggs on 11th May and first young on 6th June.

Non-breeders remained on the coast with June gatherings of 50 at Longhoughton Steel and Hauxley and smaller groups elsewhere. Coastal numbers rose quickly during July and by mid-August 1,249 were at Lindisfarne, over 100 roosted at Cresswell and 85 were at Newbiggin.

Autumn passage and the arrival of wintering birds led to further increases to provide late peaks of 1,543 at Lindisfarne in October with 1,201 being noted along the rest of the coastline in November.

**Avocet,** *Recurvirostra avosetta*
A rare visitor.

The only record involved a single which was in the Newton Pool and Embleton areas between 30th March–3rd April.

**Little Ringed Plover,** *Charadrius dubius*
A rare casual breeding species and passage visitor.

Spring arrival was marked in April by two at Cresswell Pond between

22nd–25th and a single at Alnmouth on 26th.

During May two continued to frequent Cresswell up to 16th, a single paused at Holywell Pond and three pairs were noted at inland sites. Two pairs went on to fledge young at one site and at another display and nest-scraping were seen although the outcome was not known. A pair was present at a third site between April-July, also suggesting attempted breeding.

Singles visited Cresswell on 25th June and the Long Nanny Burn on 14th July while during August individuals were at Druridge Pools on 1st and Cresswell between 10th-16th.

**Ringed Plover,** *Charadrius hiaticula*
A well-represented breeding species and a passage and winter visitor.

During the first quarter small parties were thinly distributed along the coast. An early peak of 257 on the January wader count included a concentration of 74 at S Alnmouth.

Pairs were displaying at one coastal breeding site by 29th February while a return to riverside and reservoir territories began during March and continued into April. During May pairs were reported from at least 35 breeding sites while typical late and strong passage N of Arctic birds produced peaks of 908 at Lindisfarne, 99 at Blyth and 71 at Boulmer.

Results from breeding areas included 12 pairs nesting on the Farne Islands, compared with eight in 1991. The first clutches were recorded on 25th April and first broods on 21st May. Only one pair was on Coquet Island, producing three clutches. The first two broods were predated, probably by Black-headed Gulls, *Larus ridibundus*, and the final clutch was lost at the incubation stage.

At a well-monitored mainland site ten-13 pairs hatched 64 young but only about 14 fledged after heavy losses from predation and abnormally high tides. The typically protracted breeding season was indicated by small young at one site on 19th August.

A return to coastal areas and influxes of passage birds quickly increased numbers in August-September with peaks of 142 at N Blyth, 70 at Druridge Bay and Boulmer and 65 at Ross. Onward passage then caused a decrease and during the final quarter a peak of 409 was noted on the December count, including 81 at S Alnmouth.

**Dotterel,** *Charadrius morinellus*
A rare and irregular passage visitor.

An autumn passage party of 12 adults and a juvenile feeding in a field near Howick on 31st August was the only record.

**Golden Plover,** *Pluvialis apricaria*
A common breeding species and an abundant passage and winter visitor.

Coastal fields and mudflats attracted the largest concentrations during the first quarter with peaks of 3,000 between Beadnell-Seahouses, 1,750 at Lindisfarne, 1,400 near St Mary's Island and 1,000 at Cresswell and Monks House.

Passage to breeding areas quickly reduced numbers during March and pairs and small parties were noted in seven inland nesting areas by the end of the month. Birds of the Northern race, *P.a. altifrons*, were on passage during April with the largest parties including 120 at Grindon, 50 at Blanchland and 40 at Wellhope Moor when local birds were settled into their high moorland breeding areas. The largest breeding concentration was at least seven pairs in a small area of the Simonsides and pairs were also reported from seven other typical localities.

Post-breeding flocks quickly moved back to the coast in July with over 300 at both Longhoughton Steel and Budle Bay. Numbers rapidly built up throughout August-September with the arrival of passage and wintering parties.

During the final quarter coastal areas again held the main populations with peaks of 6,050 at Lindisfarne, 2,800 near St Mary's Island, 2,400 between Beadnell-Seahouses and 1,100 at Newton. Smaller flocks frequented other coastal areas and during November parties of up to 200 remained inland at Grindon and Caistron.

**Grey Plover,** *Pluvialis squatarola*
A well-represented passage and winter visitor.

Once again Lindisfarne was the favoured area with an early peak of 1,530 in February. The rest of the wintering population was scattered thinly along the rest of the coastline with a first quarter maximum of 124 noted during January, including 40 in the regular winter haunt at Longhoughton Steel.

Numbers fell during March-April with the departure of local wintering birds for Arctic breeding grounds. Passage of flocks which had wintered further S was evident during May with a peak of 1,746 at Lindisfarne and small groups in 12 other localities, many in breeding plumage.

As usual a few non-breeders remained in summer with one-two noted at St Mary's Island, Newbiggin and Longhoughton Steel. Return passage began in July with up to 60 gathered in Budle Bay by the end of the month and continued during August.

Large-scale arrivals occurred in September providing a peak of 2,000 at Lindisfarne although, typically, they remained scarce along the rest of the coast. Onward passage led to a decline in October leaving 450 at Lindisfarne in November when a total of only 50 were in other localities on the monthly count.

**Lapwing,** *Vanellus vanellus*
A common breeding species, abundant as a passage and winter visitor.

Coastal fields and lowland arable and pastureland held the main concentrations in the first quarter. The January wader count recorded a peak of 4,170 along the coast while 2,525 was the maximum count for Lindisfarne.

In mild conditions, significant numbers remained inland with, for example, 570 at Caistron, 500 at Heddon, 450 at Hexham and 400 at Derwent in January-February.

A return to breeding areas began in late February with birds displaying in the Irthing Valley and at Chipchase. During March larger-scale movement inland occurred with pairs established in many breeding areas as the month ended.

Breeding followed in many typical habitats with young being noted from 2nd May. However, in the SE breeding numbers were reported to be well down on previous years.

Post-breeding flocking occurred in July with concentrations of 1,000 at Caistron, 520 at Derwent, 400 at Grindon and 200-250 at Wallsend Swallow and Arcot ponds and Long Nanny Burn and smaller concentrations in many other localities.

Numbers continued to build up during autumn and early winter when most flocks were again concentrated in coastal areas. They provided a late peak of 6,800 on the December wader count although some large flocks were still inland with 250-400 at Hexham, Grindon, Derwent, Caistron, Scotswood and Longbenton. During December 3,146 were wintering at Lindisfarne.

**Knot,** *Calidris canutus*
An abundant passage and winter visitor.

The favoured wintering areas at Lindisfarne and the St Mary's Island-Newbiggin stretch of coast again attracted most of the population.

During the first quarter there were peaks of 2,000 feeding alongside Holy Island causeway on 26th January, 450 frequented St Mary's Island in February with birds from these main concentrations probably accounting for sightings of parties of 60-160 at Seahouses and Blyth. A roost at St Cuthbert's Island, Lindisfarne, still held 500 in mid-April before passage to Arctic breeding grounds took most birds from the area. Only one-four remained at four localities during May and none were seen in June.

Typical early return passage, including birds still in red breeding plumage, occurred in July with 34 in Budle Bay and ten at Cresswell. Numbers increased during August while in September up to 400 were at St Mary's Island.

Many observers considered them particularly scarce during the final quarter. 1,250 were recorded in the St Cuthbert's Island roost in October but only 618 were counted at Lindisfarne in December when about 200 were wintering at St Mary's Island.

**Sanderling,** *Calidris alba*
A well-represented passage and winter visitor.

Sandy tideline feeding areas attracted small gatherings during the first quarter with peaks of 77 in the Tyne estuary, 70 at Lindisfarne and 50 at Berwick Little Beach.

Movement back to breeding areas during April led to a decline of local wintering flocks although in May continued passage led to peaks of 100 at Lindisfarne, 68 at Long Nanny Burn and 47 at Hauxley.

Some non-breeders remained in summer with one-two present at four localities.

Return passage began in late July with parties of 30 at Hauxley and Long Nanny Burn. Arrivals increased in August to bring 90 to Cambois and smaller parties to several other localities while in September 114 were at Lindisfarne and 79 remained at Cambois.

During the final quarter groups were settled in several typical feeding areas but numbers were generally low with, for example, a late peak of 216 noted on the November wader count when most were in Druridge Bay.

Inland records are rare but a single adult was at Hallington on 23rd August.

**Semipalmated Sandpiper,** *Calidris pusilla*
An extremely rare visitor.

One was discovered on Knoxes Reef, Farne Islands, on 15th June and remained until 18th. It provided the first county record and a full account appears as a special paper at the end of this report.

**Little Stint,** *Calidris minuta*
An uncommon passage visitor.
Spring passage occurred in May with four at the Long Nanny Burn on 19th, two at Druridge Bay on 24th and singles noted during this period at Holywell Pond and the Farne Islands. Two were also at Druridge on 22nd June.
Return movement started in July with a single at Budle Bay on 18th and six at Long Nanny Burn on 30th. As usual, numbers increased in August-September with one-three at nine coastal localities and singles at Holywell Pond on two dates.
The final records were in early October with singles at the Farne Islands on 1st and Druridge Pools on 3rd.

**Temminck's Stint,** *Calidris temminckii*
A rare passage visitor.
A single at Holywell Pond from 2nd-3rd September was the only record.

**Pectoral Sandpiper,** *Calidris melanotos*
A rare visitor.
There were two records: an adult in the Cresswell, Bell's Pond and Druridge area from 5th-26th August and a juvenile at Druridge Pools from 3rd-6th October.

**Curlew Sandpiper,** *Calidris ferruginea*
An uncommon passage visitor.
Very light spring passage in May provided two at Druridge Pools from 21st-26th and in Blyth estuary on 28th. A single followed at Long Nanny Burn on 5th June.
Return movement was evident in late July with a single at Ladyburn Lake, an adult in breeding plumage and an immature at Cresswell Pond and two at Long Nanny Burn, all between 26th-30th. Passage increased in August with one-three at nine coastal localities and a single inland at Derwent.
Onward passage quickly led to a reduction in numbers with only a single at St Mary's Island in September and individuals at Beadnell and Holy Island in October. The final sighting was a single at N Blyth on 5th November.

**Purple Sandpiper,** *Calidris maritima*
A well-represented passage and winter visitor.
The usual rocky habitats held most of the population with an early peak of 400 on the January wader count, including 200 in the favoured Bamburgh area. The only other large concentrations were 60 in the Beadnell-Seahouses and the Chevington Burn-Amble stretches.
Numbers declined in March-April with the departure of wintering flocks although 260 remained on the Farne Islands throughout April-May. 80 also remained at Newbiggin into early May.
There were no June records. Return passage began on 3rd-4th July with two-four noted at four localities. Numbers at mainland areas remained very low during August-September with only one-six noted in five areas. However, flocks on the Farne Islands built up to over 500 by late September.
Small-scale arrivals were evident during the final quarter to produce a late peak of 308 on the December wader count when most were again concentrated between Budle Point-Seahouses.

A colour-ringed bird at the Farne Islands in October had been marked at Vlieland, Netherlands, in September 1986.

**Dunlin,** *Calidris alpina*
An uncommon breeding species but an abundant passage and winter visitor.

Numbers were lower than usual in most areas during the first quarter with early peaks of 5,800 at Lindisfarne and 1,739 along the rest of the coastline in January.

Flocks quickly declined with onward passage and some very early movement back towards upland breeding areas was noted. For example, a single was at Whittle Dene in late January, two were at Grindon Lough in February while during March pairs and small parties were noted in four inland areas. Breeding followed with at least eight pairs being found in three typical upland sites in the SW.

As local pairs were settling into breeding territories, passage N continued and during May flocks of 400 were at Budle Bay, 73 at Boulmer and 50 at Blyth.

As usual a few non-breeders summered on the coast with one-eight noted at five localities in June.

Return passage commenced in July with 400 at a high tide roost in Budle Bay by 31st. Continued influxes provided counts of 150 at Boulmer and Blyth in late August with numbers continuing to build up into the final quarter.

During September 1,845 were back at Lindisfarne rising to a late peak of 3,178 by December. Along the rest of the coastline there was a late peak of 764 on the December count, significantly down on the 2,238 in the same period of 1991.

**Ruff,** *Philomachus pugnax*
A well-represented passage visitor, uncommon in winter.

A single wintering bird was at Lindisfarne in January. Spring passage began in April with two at Druridge Bay on 8th followed by five at the same locality on 22nd. Singles passed through three other localities.

Movement N increased in May with peaks of nine at Druridge, seven at Backworth, five over the Farne Islands and at Long Nanny Burn and one-four at three other localities. One-two lingered at Druridge Bay and Long Nanny Burn in June.

Very light return passage started in July with one-three on seven dates at Druridge Bay before increasing sharply in August when the species proved one of the highlights of late summer. There was an inland peak of 30 at Hallington and ten at Whittle Dene late in the month but most birds were on the coast. For example, 75 in the Druridge Pools-Cresswell Pond area on 30th August built up to a peak of 146 by 13th September. Smaller parties at Hauxley, Newbiggin and Castle Island may have involved some of these birds. Groups of up to 12 passed through the Farne Islands and there was a peak of seven at Lindisfarne.

By mid-October most had moved on although up to 23 remained in the Druridge Bay area up to 17th but few were seen elsewhere. During November up to 14 remained at Cresswell and Linton while as late as 20th December 12 lingered in the same areas.

**SPRING RUFFS – display is always a big attraction**

**Jack Snipe,** *Lymnocryptes minimus*
An uncommon passage and winter visitor.
    Typical wetland areas at Whittle Dene, Hallington, Broomlee Lough, Cresswell Pond, Druridge Pools and Warkworth Gut held one-two wintering birds between January-April. One-two were noted at Holywell Pond and Druridge Pools during May and a single was still at the latter locality on 14th June.
    None were seen in July but return passage was evident from 9th September with a single at Lindisfarne followed later in the month with other individuals at Druridge Pools, Hauxley, Farne Islands and Cocklawburn. Arrivals increased in October with one-two in seven coastal localities with other individuals inland at Whittle Dene and Stamfordham.
    During November-December singles were again settled in winter haunts at Hauxley and Colt Crag and individuals were also seen at Newbiggin and Cramlington.

**Snipe,** *Gallinago gallinago*
A common breeding species and a passage and winter visitor.
    They were unusually scarce during the first quarter with feeding concentrations noted in only a few favoured areas. These provided peaks of 36 at Whittle Dene, 13 at Druridge Pools, 12 at Holywell Pond and only one-two reported from other widespread localities.
    Spring passage may have been involved in a count of 32 at Druridge

Pools on 14th April while during April-May one-three on nine dates at the Farne Islands probably involved emigrants.

During April-June local birds were settled in breeding areas with 'drumming' display being recorded from widespread localities.

The usual early post-breeding gatherings began to form in July with 40 at Druridge Pools by 8th and ten at Big Waters on 20th increased to 24 by early August. Other larger groups during August-September involved 19 at Whittle Dene, 18 feeding in a flooded field at Cramlington and 11 at Caistron.

The arrival of winter visitors then increased numbers to provide peaks during the final quarter of 40 at Big Waters, 26 at Caistron and the Coquet estuary, 20 at Newbiggin, Druridge Pools and Hauxley Reserve and 13 on the Breamish at Brandon.

**Woodcock,** *Scolopax rusticola*
A well-represented breeding species and a passage and winter visitor.

Singles were wintering in several widespread localities in January-February and an individual first noted in late 1991 regularly feeding on grassed areas between office buildings at Longbenton remained in the area.

'Roding' display occurred over many suitable woodland breeding areas from March while emigration was indicated by a single on the Farne Islands on 7th April. During the breeding season birds continued to be reported from many typical inland localities.

The first indication of passage was a single at Blyth in late September while in the first half of October light passage was evident with one-two in seven coastal and six inland localities. Arrival then increased with two-three daily at Holy Island between 25th-30th and similar movements being noted on the Farne Islands.

Further arrivals occurred in November with six at Newbiggin on 9th and one-three at 17 other coastal localities, including tired birds coming into gardens. One-three were also noted in four inland areas offering good feeding. Extremely cold weather in December led to some moving to the coast with four being flushed from bushes at Cresswell on 31st.

**Black-tailed Godwit,** *Limosa limosa*
An uncommon passage visitor, rare in winter.

Up to four birds spent January-March in the regular wintering area at Budle Bay. Spring passage in April then produced parties of nine at Druridge Pools on 21st when three flew N at the Farne Islands. During late April one-six were also seen at six other coastal localities and inland singles visited Gosforth Park and Grindon Lough.

Passage continued in May with maximum counts of 15 at Druridge Pools and nine at Hauxley and further singles at Holywell Pond and Grindon Lough.

There were no June records and early return passage commenced in late July with 12 feeding at Long Nanny Burn on 30th when eight flew S at Seaton Sluice. One-three were noted at three other coastal localities. During August 11 flying S at the Farne Islands on 4th was the largest group. Elsewhere passage in August-September was on a lighter scale than in recent years with one-four at ten coastal sites and peaks of four inland at Big Waters and one-two at Arcot and Holywell ponds.

During October-December up to three were wintering at Lindisfarne and a single was at Boulmer.

**Bar-tailed Godwit,** *Limosa lapponica*
A common passage and winter visitor.
   Most were concentrated at Lindisfarne with an early peak of 3,590 during January. Numbers were characteristically low elsewhere with only 74 recorded along the rest of the coastline during February.
   Departure for breeding grounds occurred in March-April, although 180 were still at Lindisfarne in early May. During June one-six summering individuals remained in six localities.
   The usual early return passage, including birds still in breeding plumage, began in July, first with small parties moving past Hauxley and Seaton Sluice and by 29th 150 were in Budle Bay. Larger arrivals occurred in August with 1,931 at Lindisfarne by 16th. Small parties were recorded right along the coast during September as they moved through the region.
   During the final quarter there was a peak of 2,202 at Lindisfarne in October when only 99 were found along the rest of the coastline.

**Whimbrel,** *Numenius phaeopus*
A well-represented passage visitor.
   The first indication of spring passage was a party of five calling over Holy Island on 19th April. They were quickly followed by one-two at four other coastal localities later in the month. During May the largest parties were ten at Boulmer on 9th and 19 at Hauxley on 20th while groups of up to seven passed through ten other coastal localities. Only four individuals were noted in June.
   They became prominent on return passage in July with parties of up to 15 at Lindisfarne, the Farne Islands and Seaton Sluice, 11 at Long Nanny Burn, eight at Cresswell and up to four in 11 other coastal areas.
   August was the peak month with 90 moving S at Blyth on 7th, 27 S at Holy Island on 9th, 13 at Hauxley on 21st and groups of up to 11 at ten other localities. Numbers fell quickly in September with groups of one-seven in five areas. The last occurred in October with eight at Hauxley on 3rd and singles at Druridge on 4th and the Farne Islands on 8th.

**Curlew,** *Numenius arquata*
A common breeding species and a passage and winter visitor.
   Coastal numbers were much lower than usual during the first quarter with an early maximum of 424 on the February wader count and a peak of 410 at Lindisfarne. Considerable numbers remained inland with a winter gathering of 125 at Caistron in January and 47 moving E at Scotswood, indicating that they had been feeding further up the Tyne.
   Many moved back into nesting areas in March providing pre-breeding gatherings in several localities, the largest involving 122 at Grindon Lough. During April-May most were displaying and settling into breeding areas. An albino near Simonburn may have been the individual noted breeding annually in the district between 1987-90.
   Small parties of non-breeders remained on the coast with up to 30 at eight localities during June.
   Post-breeding flocks formed from July with movement back to the coast providing gatherings of 233 at Longhoughton Steel, 150 at Annstead, 120 at Long Nanny Burn and 100 at Lindisfarne. Smaller flocks frequented many other localities during August-September with numbers gradually building up with winter arrival.

During October-December there were peaks of 660 at Lindisfarne, 600 on the Farne Islands with other gatherings including 146 at Long Nanny Burn. Again many remained inland with 97 on moorland at Holystone and 15 near Wooler in November.

**Spotted Redshank,** *Tringa erythropus*
An uncommon passage visitor, rare in winter.

A wintering individual discovered at Boulmer in December 1991 remained during January.

Typical light spring passage produced singles at Big Waters and Wallsend Swallow Pond between 20th-24th April while during May two visited the Long Nanny Burn and singles were also noted at Druridge Pools and Hauxley Reserve. Another was regularly in the Druridge Pools area in June, remaining until at least 6th July.

Return movements commenced in August with one-two at five coastal localities and inland at Pegswood, Holywell and Arcot ponds, Big Waters and Whittle Dene. September was the main passage period with peaks of five at Hauxley Reserve, four at Cresswell Pond and one-two at eight other localities. The final record was a single at Druridge Pools on 3rd October.

**Redshank,** *Tringa totanus*
A common breeding species and a passage and winter visitor.

During the first quarter parties were spread along the coast with a peak of 919 noted on the March wader count. However, roosts provided the opportunity for other accurate counts with as many as 500 using Blyth staithes and 227 at the old outdoor swimming pool at Tynemouth. As usual, very small numbers remained inland with eight on the Tyne between Newburn-Wylam in January.

Movement away from the region and the drift of local birds back to inland breeding areas occurred in late March and a pair was displaying at Arcot Pond by 4th April. Breeding followed in many suitable damp pasture, riverside and other wetland habitats, including ten pairs at Caistron where the first young were noted on 20th June.

The usual quick return to coastal haunts occurred in July with 130 at Hauxley by 15th and 200 at Lindisfarne on 30th. Numbers continued to increase with 1,265 at Lindisfarne in mid-August when over 300 were again at Blyth. 60 were at the Farne Islands in September.

Arrival of more passage and wintering birds increased the Lindisfarne population to a late peak of 1,725 in October when roost gatherings included 425 at Blyth staithes and 100 at Seaton Sluice. A late peak of 897 was found along the coast on the November wader count and 200 frequented the Tweed estuary in December. A few remained inland with five at Close House and one-three at Killingworth Lake and Hallington.

**Greenshank,** *Tringa nebularia*
A well-represented passage visitor, rare in winter.

Inland wintering records are extremely rare. However, a single was found on the Till near Chillingham on 6th March, well before spring passage could be expected, and may have wintered in the area.

Spring passage commenced in April and involved singles at Budle Bay and Prestwick Carr between 21st-25th followed by one-two at nine coastal localities in May and a single at Castle Island in early June.

Return passage began in mid-July with five feeding in Budle Bay on 18th, nine at Lindisfarne on 30th and one-two at four other coastal localities and inland at Colt Crag, Grindon Lough, Powburn, Holystone and Scotswood.

Movements increased in August with sightings in many widespread localities. There were coastal peaks of six at Druridge Pools and Foxton and inland at Hallington. Five also frequented Cresswell and one-three were at 18 other areas. Strong passage continued in September with a peak of 11 at Lindisfarne, six inland at Powburn and one-four at 13 other localities. The last record was from the Farne Islands on 27th September.

**Green Sandpiper,** *Tringa ochropus*
An uncommon passage visitor, rare in winter.

The first spring passage bird was at Holywell Pond on 4th April and was followed by two at Prestwick Carr on 22nd and another at Holywell on 23rd. During May one-two were noted on three dates at Hauxley Reserve and in June singles were at Derwent, Holywell Pond and Druridge Pools.

The usual mid-summer return movements began in July with two on the Breamish at Powburn and singles in five other localities. During August seven were at Powburn, four at Arcot Pond and up to three frequented at least 13 other localities. Passage continued in September with three at Barrasford, two at Medburn, Hallington and Castle Island and singles at eight ponds. Most departed by the end of September with singles lingering in October at Hallington on 19th and at Barrasford until 31st.

During late December singles were wintering on the Breamish between Powburn and Beanley and on the Till at Ewart Park.

**Wood Sandpiper,** *Tringa glareola*
An uncommon passage visitor.

Typically light spring passage was evident from 5th May producing one-two at Cresswell Pond, Druridge Pools, Hauxley Reserve and on the Farne Islands. In June singles were at Holywell Pond and again at Druridge Pools.

Returning individuals were found from 7th August at Hauxley Reserve followed by others at Holywell and Arcot ponds and Colt Crag. During September singles visited Holywell and Bell's ponds and Holy Island Lough. In September two at Holywell Pond on 1st and a single at Hauxley Reserve on 2nd-3rd were the last to be noted.

**Common Sandpiper,** *Actitus hypoleucos*
A common summer and passage visitor, rare in winter.

Arrival in April started with a single on the Tyne at Bywell on 16th and by the end of the month they were back in at least eight inland localities with a maximum of nine at Whittle Dene.

Many more arrived in early May when passage was evident along the coast with one-three in 18 localities while inland pairs were displaying and settled in at least 24 typical waterside localities. Surveys in July showed 23 birds on the Tyne between Corbridge-Hexham and at least four pairs present in a two kms. stretch of the Breamish at Powburn.

The very rapid departure from breeding haunts was shown in late July with 13 at the Long Nanny Burn and nine at Big Waters while in early August ten were at Druridge Pools, eight at Castle Island and seven at Cresswell Pond. During this period one-five were noted at the Farne Islands on 33 dates.

Onward passage meant that most had departed by September when only two-four were seen at three coastal localities and singles in 12 other areas. The final record was a single at North Shields Fish Quay on 1st October.

**Turnstone,** *Arenaria interpres*
A common passage and winter visitor.

Abundant feeding provided by rocky habitats again held most of the population during the first quarter. There was a peak of 1,051 along the coast during the March wader count, including 268 between Budle Point-Seahouses. The largest roost involved 150 at Blyth staithes.

Numbers decreased in April, although 100 were roosting at the Farne Islands late in the month. Further departures for breeding grounds then occurred and most had gone by the end of May.

Small parties of non-breeders remained with 25 at Hauxley, ten at Cresswell, and up to five at St Mary's Island, Longhoughton Steel and the Farne Islands in June.

Very light return passage was evident in July while in August 153 were at Lindisfarne, 145 were back between Budle Point-Seahouses and 60 were at Blyth. Fresh waves of arrivals led to a continued build-up in numbers during September when by far the biggest count was over 1,000 roosting on the Farne Islands.

During the final quarter there was a peak of 1,121 on the November count and the Farne Islands roost had reduced to 100-150 birds. The usual roost at Blyth staithes increased from 138 in November to 296 by late December.

Inland records are rare but two were seen flying NW over Whittle Dene on 16th February.

**Red-necked Phalarope,** *Phalaropus lobatus*
A rare visitor.

A juvenile landed briefly in The Kettle, Farne Islands, before flying off N on 30th August. It was the first county record since August 1986 when three-five were at Druridge Pools.

**Pomarine Skua,** *Stercorarius pomarinus*
An uncommon passage visitor, mainly in autumn.

The only late spring record involved a sub-adult N off Seaton Sluice on 4th June while in July two adults also moved N from the same locality on 2nd and a single was off the Farnes on 29th.

More regular return passage commenced in early August with one-eight at eight localities while during September one-three were noted from six seawatching sites.

None of this gave any hint of the spectacular events to follow in October. The month began with one-two between 3rd-6th but on the 9th, as a cold front moved down the North Sea, unprecedented numbers appeared. At least 488 flew S at the Farne Islands in groups of ten-30. Passage went on throughout the day with 80%-90% being adults. Also on 9th, 56 off Holy Island included a flock of 38. As the weather improved these flocks appeared to turn and head back N giving the spectacular passage of following days.

On 10th observers right along the coast enjoyed the wonderful sight of large flocks moving N and counts undoubtedly involved many of the same groups. The highest figure was 250 off Newbiggin in six hours with 30-160 being noted from five other localities between St Mary's Island and Holy Island. Passage N continued on 11th with 218 at Seaton Sluice, including a flock of 65, 170 at St Mary's Island, 132 at Hauxley while 80 in 1½ hours at Newbiggin, included a loose party of 50. The only indication of hunting or feeding involved two at Newbiggin which attacked Kittiwakes, *Rissa tridactyla*. One skua twice pulled down a Kittiwake into the sea before it managed to escape.

**POMARINE SKUAS – for many the highlight of the year**   S. SEXTON. 92...

Lighter passage continued on 12th with 95 N at Seaton Sluice while on 15th 78 off the Farnes included a flock of 70. Numbers then fell further with 28 off St Mary's Island on 16th while on 20th another 75 at the Farnes ended the huge passage.

It will never be known just how many birds were involved but it was clearly on a very much greater scale than the last big influx on October 1985. On a historical note, it was interesting that the 1992 spectacle occurred exactly 100 years after a big influx detailed by Bolam in *The Birds of Northumberland and the Eastern Borders* (1912).

**Arctic Skua,** *Stercorarius parasiticus*
A common passage visitor, uncommon in winter.

Typical light spring passage began on 15th March when a single chased Kittiwakes, *Rissa tridactyla*, off Boulmer. Another was off the Farne Islands on 15th April. During May only two were noted off Holy Island and three off the Farnes while in June 12 individuals were seen between Newbiggin and Seahouses.

Passage increased in July with 74 individuals being noted along the coast while in August there were peaks at the Farnes of 55 S on 14th and 48 in one hour on 27th. Elsewhere there was a monthly total of 189 S and 95 N at Seaton Sluice and 30 per hour at Newbiggin on 30th. During September 160 flew S and 129 N at Seaton Sluice with many sightings of individuals and small groups from other seawatching localities.

Lighter movement N continued in early October, particularly on 4th, with ten-15 seen from Seaton Sluice, Newbiggin and Hauxley. Numbers quickly dwindled with the last involving singles off Holy Island on 26th and 28th.

Inland records are rare but on 13th August a juvenile flew W at Whittle Dene. On 6th September a party of five arrived high from W at Newbiggin, perhaps indicating overland movement.

**Long-tailed Skua,** *Stercorarius longicaudatus*
Normally a rare passage visitor.

After the unprecedented influxes of 1991, a return to more normal figures occurred with only five records during late summer.

An adult was at Holy Island on 17th July and a juvenile was found dead on Coquet Island on 26th August. In September adults flew S at Seaton Sluice on 4th, Cocklawburn on 6th and at the Farne Islands on 23rd.

**Great Skua,** *Stercorarius skua*
A well-represented passage visitor, rare in winter.

Very light spring passage was evident with a single off the Farne Islands on 28th April and another N at Newbiggin on 23rd May. During June sightings of singles flying N at Seaton Sluice, Newbiggin and Hauxley on 19th may have involved the same individual. Other individuals passed N at the Farne Islands on 21st and at Seaton Sluice on 30th.

As usual, passage increased in mid-summer with 23 noted off Hauxley on 7th July and one-four were seen during the month from four other seawatching localities. During August light passage continued and one killed a young Shelduck, *Tadorna tadorna*, at Fenham-le-Moor on 1st. However, the most significant date was the 30th when nine moved N at Newbiggin, six were seen from Hauxley and three at Boulmer.

Passage N increased during September with 15 at Hauxley on 6th, 22 at the Farne Islands on 22nd and 12 at Hauxley and ten at Newbiggin on 26th., An injured bird was at Druridge Pools on 28th. Movements continued in October with peak counts of 26 at Newbiggin on 3rd and Seaton Sluice on 4th followed by ten at St Mary's Island on 10th. Sightings then quickly dwindled in November, involving only a single off the Farne Islands on 12th and four N at Newbiggin on 15th.

**Mediterranean Gull,** *Larus melanocephalus*
An uncommon winter and passage visitor.

The regular adult at Craster, first noted in 1984, was present from the start of the year until 3rd March. Another adult was noted between January-March in the Tyne Estuary-Seaton Sluice area and a second-winter bird at Druridge Pools on 27th February. In April an adult in full breeding plumage was at Hauxley Reserve on 12th and a first-summer bird at Coquet Island on 26th.

The first returning bird was the regular adult noted at Embleton Burn and Craster on 18th July and was subsequently seen on a number of dates until December. This may have been the adult at Newton Haven on 3rd August. Other adults were at Big Waters on 25th July, Seaton Sluice on 26th, Druridge Pools on 16th September, North Shields and Cambois on 14th November and N Blyth on 19th December. A second-summer bird with a metal ring was at Seaton Sluice on 26th July and a second-winter individual at Druridge Pools on 25th September.

There were three inland records in September, coinciding with a major influx of Black-headed, *L. ridibundus*, and Common gulls, *L. canus*: an adult was at Colt Crag on 22nd and an adult and a second-winter bird at Scotswood on 24th.

**Little Gull,** *Larus minutus*
A well-represented passage visitor, uncommon in winter.

A first-winter bird at Stag Rocks on 2nd February was the first sighting of the year followed by an adult at Holywell on 24th. An adult was at Newbiggin on 15th March with another at North Shields Fish Quay from 26th March-28th April. Other early spring records included single first-winter birds on Coquet Island and in Budle Bay on 27th-29th March and at least two birds in the Druridge area from 12th-25th April.

Numbers increased in May with five at Cresswell Pond on 27th and one-four at six other coastal localities. Peak counts were in June-July with 14 at Druridge Pools, 12 at Cresswell Pond and nine at Castle Island. Ageing data showed that 100% of the birds were in their first-summer in May, 75% in June and 60% in July.

Autumn passage was marked by eight flying S at Seaton Sluice on 7th July, six (including five juveniles) flying N at Newbiggin on 30th August and a total of eight flying S and eight N at Seaton Sluice between 16th-21st September. In October, one was at Ladyburn Lake on 3rd, a juvenile flew N at Newbiggin on 11th, an adult was at Cambois on 24th and an immature in Holy Island Harbour on 25th.

## TABLE 1. MAXIMUM COUNTS AT GULL ROOSTS DURING 1992

| | JAN | FEB | MAR | APR | JUL | AUG | SEPT | OCT | NOV | DEC |
|---|---|---|---|---|---|---|---|---|---|---|
| **BLACK-HEADED GULL** | | | | | | | | | | |
| Scotswood | 8,000 | 3,400 | 300 | - | 2,950 | 3,300 | 3,300 | 3,600 | 4,700 | 130 |
| Derwent | 1,300 | 800 | 2,700 | 1,500 | - | 750 | 800 | 6,000 | 1,400 | 650 |
| Hallington | 5,600 | 750 | 3,900 | 1,500 | 210 | 800 | 17,000 | 5,000 | 1,600 | - |
| Colt Crag | - | - | 670 | 1,200 | 1,050 | 620 | 4,400 | - | - | 250 |
| Grindon | - | 65 | 6,200 | 4,300 | 105 | - | - | - | - | - |
| Caistron | 330 | 200 | 7,500 | 6,000 | - | - | - | - | 345 | 330 |
| Budle Bay | 350 | 600 | 900 | 14 | 570 | - | 2,400 | 3,500 | 990 | - |
| **COMMON GULL** | | | | | | | | | | |
| Derwent | 7,300 | 4,800 | 8,500 | 10,500 | 1,170 | 4,300 | 5,600 | 9,500 | 11,500 | 3,200 |
| Hallington | 4,900 | 22,000 | 41,000 | 27,000 | 135 | 2,400 | 14,000 | 700 | 28,000 | 320 |
| Colt Crag | - | - | - | 350 | 2,950 | 5,200 | 28,000 | 15,000 | 980 | 2,950 |
| Broomlee | - | 220 | 740 | 75 | - | 86 | - | 540 | 1,020 | - |
| Budle Bay | 1,550 | 3,100 | 480 | 90 | 930 | - | 12,000 | 8,500 | 4,700 | - |
| Tyne Estuary (day) | 1,100 | 150 | 1,800 | - | - | - | - | - | 100 | 400 |
| **HERRING GULL** | | | | | | | | | | |
| Scotswood | 350 | 470 | 240 | - | 480 | 460 | 530 | 140 | 520 | - |
| Derwent | 3,600 | 210 | - | 210 | - | - | 116 | - | 270 | 300 |
| Hallington | 710 | 120 | 130 | - | - | 60 | 60 | 75 | 520 | 550 |
| Tyne Estuary (day) | 1,050 | 1,300 | 600 | 550 | 450 | - | 700 | 500 | 1,150 | 1,200 |

**Sabine's Gull,** *Larus sabini*
A rare visitor.
A first-summer bird flew N at Seaton Sluice on 1st August and an adult, resting on rocks at Newbiggin on 6th September, provided excellent views.

**Black-headed Gull,** *Larus ridibundus*
A common breeding species, abundant as a passage and winter visitor.
Details of maximum counts at regular roost sites are given in Table 1. The largest roost in the first two months was at Scotswood with 8,000 present on 6th January. Large mixed roosts, including some Common Gulls, *L. canus*, were noted at Tynemouth with 8,500 on 14th January and 6,000 on 8th February. Elsewhere on or near the coast, gatherings of 1,000-2,000 were seen at Blyth, Seaton Sluice, Boulmer and Holywell in January.
By the end of March, the main concentrations were at or near breeding sites with 7,500 at Caistron and 6,200 at Grindon. In April, movement over the Wark Forest was noted on several dates with 100 flying NW on 25th.

Breeding data was as follows:

|  | no. pairs/nests (1991 in brackets) |  | no. young fledged where known |
|---|---|---|---|
| Coquet Island | 4,024 | (2,771) | 2,696 |
| Caistron | 1,450 | (500+) | 700 |
| Moorland Sites (SW) | 1,427 | (1,392) | 330 |
| Colt Crag | 320 | (250) | 540 |
| Farne Islands | 107 | (181) | - |
| Holy Island Lough | 100 | (0) | - |
| Newton Pool | 52 | (17) | - |
| Darden Lough | 40-50 | (40) | - |
| Beanley | 40 | (0) | - |

The poor breeding success at the main moorland site led to 930 roosting at Scotswood by 28th June but the other two main sites had a typical year with 0.5-0.7 young fledged per breeding pair.
By the end of July 2,950 were roosting at Scotswood but numbers were generally unexceptional until mid-September when a roost of 17,000 at Hallington was the highest recorded inland so far. At the same time, numbers increased on the Farne Islands and counts of 800-4,400 were recorded from three inland and two coastal sites. Numbers then steadily declined until the cold weather of December.
An extreme leucistic, almost albinistic bird, was at Holywell Pond on 21st October and Whittle Dene on 28th.
One ringed as a pullus at Randaberg, Norway, in June 1989 was found dead at Seaton Delaval in January, again demonstrating the Scandinavian origins of many of our wintering gulls.

**Common Gull,** *Larus canus*
An abundant passage and winter visitor and uncommon breeding species.
Details of maximum counts at regular roost sites are given in Table 1. The spring influx started early with 22,000 roosting at Hallington by 15th February. This roost increased further in March to 34,000 on 11th and 41,000 on 24th. Moderate N-NE passage was noted at Ordley on 28th-30th March

and at Forestburn Gate on 30th and the Hallington roost had declined to 27,000 by 1st April. Massive N movement occurred on 8th at Ordley when steady E movement was seen at Scotchcoultard and Thirlwall. Coastal passage was indicated by 500 at Newbiggin on 5th and, at Seaton Sluice, 65 per hour flying N on 3rd and 150 on 16th. The Hallington roost held only 5,300 birds by 13th and 230 by 24th.

The breeding population continues to increase with 17 pairs attempting to breed (14 in 1991) of which 12 were successful, raising 22 young at seven sites. Three pairs bred on moorland and nine in lowland gravel pits and quarries. The liking of the species for industrial sites was shown by a further five pairs immediately moving in to breed at a new opencast site on the SW moors. Some young did not fledge until early August.

Return movement of first-summer birds occurred in mid-July when roosts of 1,020-1,170 at Derwent and Colt Crag from 9th-15th comprised 85% birds of this age. The return of adults was leisurely in early autumn with 2,950 at Colt Crag by 29th July and 5,200 by 28th August. The main influx occurred in mid-September with 28,000 roosting at Colt Crag on 22nd and 12,000-14,000 at Hallington and Budle Bay from 17th-19th. Numbers declined in October before a second wave of immigration in November brought 28,000 to Hallington on 24th. Freezing weather in December reduced numbers inland sharply with the highest count being only 3,200 at Derwent on ice floes on 21st.

More attention was paid to identifying individuals showing characteristics of the race *L. c. heinei*. All records are of adults. At the peak of the spring migration on 24th March, 35 in a pre-roost gathering of 11,000 were attributed to this race. In the autumn, further records were of three at Hallington on 19th September and at Colt Crag on 22nd, 16 out of a flock of 4,000 at Hallington on 19th October, seven at Broomlee on 14th November and one at Caistron on 22nd.

**Lesser Black-backed Gull,** *Larus fuscus graellsi*
A common breeding species and passage visitor. Uncommon in winter. Numbers at the main breeding site were low in 1992.

The wintering population increased with one-four in 14 localities in January-February. Numbers rose in spring with the roost at Broomlee holding 18 on 6th March, 41 on 21st and 35 on 18th April. 30 were at Budle Bay on 19th and 13 flew N at Seaton Sluice in the first half of April.

Breeding data was as follows:

|  | no. pairs/nests (1991 in brackets) |  | no. young fledged where known |
|---|---|---|---|
| Farne Islands | 656 | (998) | - |
| Coquet Island | 56 | (20) | 5 |
| Moorland Sites (SW) | 6 | (11) | 6 |
| Holburn Moss | 2 | (0) | - |
| Caistron | 1 | (0) | 0 |
| Westerhope | 1 | (1) | - |

The Farne Islands count is based on the assumption in the last Operation Seafarer survey that 80% of the pairs of large gulls were this species. Breeding continued on roof-tops in Newcastle city centre.

Large mid-summer gatherings included 140 roosting at Scotswood on

28th June, 106 at the Long Nanny on 10th and 58 in Budle Bay on 21st. Numbers increased in July with 185 at Scotswood on 13th and 130 at the Long Nanny on 11th. The large autumn roost at Caistron again occurred with 54 there on 18th September and 347 on 18th October. The last significant numbers reported were 12-14 in Budle Bay on 1st-3rd November and seven at Grindon on 14th. One-two were noted in December from seven localities.

Two birds showing the characteristics of the race, *L.f. intermedius*, were at Scotswood on 6th January and 10th February with singles at Amble on 2nd February, Holy Island on 10th October, Hallington on 19th October and 24th November, Big Waters on 16th December and Derwent on 21st. Characteristics of the race, *L.f. fuscus*, were shown by single adults at Scotswood on 21st January and Derwent on 17th October.

One ringed as a pullus on the SW moors in July 1982 was killed by a falconer's hawk at Dunblane, Scotland in July.

**Herring Gull,** *Larus argentatus*
A well-represented resident breeding species and common passage and winter visitor.

Details of maximum counts at regular inland roost sites and in the Tyne Estuary are given in Table 1. An early peak was very marked with 3,600 roosting at Derwent on 10th January and 710 at Hallington on 11th. The highest counts on or near the coast in the first three months were 1,700 at Holywell on 31st January, 1,300 in the Tyne Estuary on 16th February and 1,000 on Coquet Island in March. A steady N movement was observed all day at Hauxley on 19th January with a peak count of 270 passing in 15 minutes.

Breeding data was as follows:

|  | no. pairs/nests (1991 in brackets) | | no. young fledged where known |
|---|---|---|---|
| Farne Islands | 164 | (250) | - |
| Needles Eye-Scot. border | 122 | (200) | - |
| Westerhope | 25 | (20) | - |
| Holburn Moss | 20-26 | (13) | 30 |
| Coquet Island | 20 | (5) | 0 |
| Moorland Sites (SW) | 1 | (4) | 0 |

The Farne Islands count is based on the assumption in the last Operation Seafarer survey that 20% of the pairs of large gulls were this species. Breeding continued on roof-tops in Newcastle city centre.

In mid-summer 720 roosted at Scotswood on 28th June. The largest concentration in autumn was on the Farne Islands where 5,000 roosted in mid-September. Large numbers remained there until early December and other gatherings were 1,200 in the Tyne Estuary on 6th December, 1,000 at Linton Pond on 29th, 600 in the Blyth Estuary on 31st October and 550 at Hallington on 23rd December.

A number of individuals showing characteristics of the race *L.a. argentatus* were reported in January-February and from September-December, including three of the Thayeri-type at Scotswood on 6th January and two of this type at North Shields from 1st-17th. A first-winter bird at North Shields Fish Quay on 24th-26th March showed characteristics of a hybrid Herring-Glaucous Gull *L. argentatus/hyperboreus*.

**Yellow-legged Gull,** *Larus cachinnans*
A rare passage visitor that inter-bred in 1992.
All records were of adults. One was at Riverside Park, Walker, on 21st March and 1st-30th April and another in Budle Bay on 10th May.
One bird on Coquet Island from 10th May-20th July was paired with a Lesser Black-backed Gull, *L. fuscus graellsi*. One clutch of three eggs was laid which failed to hatch.
There was a significant post-breeding influx. Singles were at Holywell on 27th July, Seahouses-Budle Bay on 31st July-2nd August, Big Waters on 12th September and Amble on 17th September.

**Iceland Gull,** *Larus glaucoides*
An uncommon passage and winter visitor.
Single first-winter and second-winter birds were reported in the Tyne Estuary from January-March. The second-winter bird may have been the same individual noted at Holywell Pond on 24th February and the first-winter bird may have wandered to Wallsend Swallow Pond on 7th March. Single adults were seen at Holywell Pond on 28th January and Wallsend Swallow Pond on 18th March.
The second-winter bird remained at North Shields until 28th April and was probably the individual at Holywell on 15th April. Indications of spring passage were provided by first-winter birds at Newbiggin on 15th March and Seaton Sluice on 27th, second-winter birds on Coquet Island on 27th March and the Farne Islands on 16th April. An adult was at Newbiggin on 5th April.
As usual autumn records were scarce. A first-summer bird flew N at Hauxley on 22nd August and an adult was at North Shields Fish Quay on 10th October. Records increased in December with a first-winter bird and an adult in the Tyne Estuary and another first-winter bird at Druridge Pools.

**Glaucous Gull,** *Larus hyperboreus*
An uncommon passage and winter visitor.
Over the first four months, at least two adults, a second-winter and five first-winter birds were reported from North Shields Fish Quay and adjacent areas. In the Druridge area, a first-winter bird and an adult were noted in January-February and a second-winter bird was at Ross on 2nd February.
There were a number of indications of spring passage. An adult was at Boulmer on 14th March. Secondwinter birds flew N at Newbiggin on 18th March and were noted on Coquet Island on three dates from 21st March-8th April and at Hadston on 23rd April. First-winter birds were in the Seaton Sluice area from 29th-31st March, on the Farne Islands on 5th and 18th-20th April and Holy Island on 20th. Less expected in our area was a first-winter bird inland which associated with Lesser Black-backed Gulls, *L. fuscus graellsi*, roosting at Broomlee on 18th April and resting on the Tyne at Wylam on 23rd April.
Passage continued into May with an adult flying N at Snab Point and a second-summer bird flying W at Cresswell Pond on 2nd. A first-summer bird was at St Mary's Island on 11th and North Shields on 16th. The last spring record was a second-summer bird on Coquet Island on 4th June.
Early autumn records comprised a first-summer and a second-winter bird in the Tyne Estuary on a number of dates from 19th July-14th August, and 6th September-10th October, a first-summer bird on the Farne Islands on 28th August, an adult at Newbiggin on 30th and another at Seahouses on

19th-20th September. This latter individual was seen on five further dates in October-November at Seahouses and the Farne Islands. An adult flew S at St Mary's Island on 17th October.

In December two adults and a first-winter bird were reported from North Shields Fish Quay.

**Great Black-backed Gull,** *Larus marinus*
A common passage and winter visitor and a rare breeding species.

Inland roosts produced the usual peak in numbers in January with 186 at Derwent on 10th and 95 at Hallington on 11th. Elsewhere during the first three months, 109 were in the Tyne Estuary on 10th January and 104 in Blyth Harbour on 22nd March.

Two pairs bred on the Farne Islands (one in 1991). The highest summer count was of 62 in the Tyne Estuary on 7th June.

Numbers increased from mid-July with 260 at Newbiggin on 30th, 104 at Seahouses on 23rd August and 100 at North Shields on 30th August. From September to the end of the year, the highest counts on the mainland coast were 504 in Blyth Harbour on 1st November, 224 in the Tyne Estuary on 5th September and 216 at Holy Island on 6th September. The largest inland roost was 120 at Derwent on 21st November. However, as in 1991, by far the largest numbers were on the Farne Islands where 2,000 were present in October-November.

**Kittiwake,** *Rissa tridactyla*
A common breeding species and abundant passage visitor.

The highest count in the first two months was 41 at North Shields on 21st February. Heavy N passage was noted on 15th March with up to 1,000 per hour at Boulmer and 400 per hour at Seaton Sluice and Newbiggin and again on 24th when 242 passed in one hour at Newbiggin. Heavy N movement was also noted in May when 500 per hour passed Seaton Sluice and Hauxley on 2nd and 590 passed Holy Island in 30 minutes on 9th.

Breeding data was as follows:

|  | no. pairs/nests (1991 in brackets) | no. young fledged where known |
|---|---|---|
| Farne Islands | 6,178 (5,743) | - |
| Needles Eye | 1,680 (1,550) | - |
| Cullernose-Howick | 588 (536) | - |
| Dunstanburgh | 564 (608) | - |
| Seahouses | 126 (80) | - |
| CWS, Newcastle | 70 (55) | 70 |
| Tynemouth Cliffs | 55 (13) | - |
| North Shields | 33 (59) | - |
| Coquet Island | 14 (1) | 9 |

Heavy N movements were noted in October with hourly rates of 258 at Snab Point on 11th, 1,500 at Seaton Sluice on 12th, 2,000 at the Farne Islands and 600 at Annstead on 20th and 700 at Seahouses on 25th. The highest count in the last two months was 27 at North Shields on 6th December.

Inland records were as follows: In March six adults were at Hallington on 11th, one at Grindon on 14th and one at Colt Crag on 18th. Five adults roosted at Hallington on 13th April. In August single adults were at Big Waters on 9th, Hallington on 25th and a juvenile was at Holywell Pond on 10th.

A bird ringed as a pullus on the Farne Islands in June 1982 was found on the Faeroes Bank in January.

**Lesser Crested Tern,** *Sterna bengalensis*
An extremely rare visitor that has inter-bred

The regular bird returned for its ninth year at the Farne Islands on 1st May and it was seen mating with a male Sandwich Tern, *S. sandvicensis*, on 13th. The first egg was laid on 25th and a second shortly after. Both young hatched on 23rd June but one was found dead on 30th. The survivor fledged on 23rd July, the adult having already departed on 21st.

The adult was also noted, prior to incubation, at the Long Nanny Burn on 19th May and, immediately after leaving the breeding site, at Newbiggin on 22nd July.

The records at the Long Nanny Burn and Farne Islands have been accepted by the British Bird Rarities Committee. The Newbiggin record is still under consideration.

**Sandwich Tern,** *Sterna sandvicensis*
A common breeding summer and passage visitor.

The first arrivals were on 18th March when one flew N at Newbiggin and one was fishing off Stag Rocks. Numbers increased in May when 600 were at Budle Point on 10th and 150-315 at Beadnell Bay, Longhoughton Steel and Cheswick Sands.

2,730 pairs bred on the Farne Islands (2,126 in 1991) with the first young flying by 4th July and 2,131 on Coquet Island (1,736 in 1991) where the first young fledged on 28th June and breeding success was reported to be good.

Departing birds were noted from 3rd July at Seaton Sluice with 600 per hour flying S. At the same locality 80 per hour flew S on 28th August and 40 per hour S on 18th September. The highest autumn mainland count was 150 at Hauxley on 13th September. Two-four birds were noted from four localities in October and the last record was one fishing off Boulmer and Alnmouth on 8th November.

The only inland record was one flying W at Cramlington on 8th August.

Nine birds ringed on the Farne Islands were reported from their W African wintering grounds, the oldest being 17 years. Three Farnes-ringed birds were found in Holland in April, presumably on return passage.

**Roseate Tern,** *Sterna dougallii*
An uncommon breeding summer and passage visitor.

First returning birds were noted in May on 4th at Coquet Island, 11th on the Farne Islands and 13th at Holy Island and Hauxley.

There was a welcome increase in numbers breeding and in productivity for this threatened species as shown below:

|  | no. pairs/nests (1991 in brackets) | | no. young fledged | first fledging |
|---|---|---|---|---|
| Coquet Island | 27-30 | (20) | 35-36 | 9th July |
| Farne Islands | 4 | (3) | 4 | 14th July |
| Lindisfarne | 0 | (1) | 0 | - |

A further three non-breeding pairs were present on Coquet Island.

Post-breeding gatherings included four at Tynemouth on 31st July and

14 at Newbiggin on 9th August. Last records were in early September with two at St Mary's Island and one on Farne Islands on 2nd and two at Holy Island on 6th.

One ringed as a pullus on Coquet Island in July 1989 was reported trapped and dead in winter quarters at Ankaful, Ghana, in January.

## Common Tern, *Sterna hirundo*

A common breeding summer and passage visitor.

First sightings were on the Farne Islands on April 21st, Coquet Island on 22nd and Hauxley on 30th.

Breeding data shown below indicates a return to former numbers after the poor season in 1991. Ten birds were at Castle Island during May.

|  | no. pairs/nests (1991 in brackets) | | no. young fledged where known |
|---|---|---|---|
| Coquet Island | 842 | (578) | 766 |
| Farne Islands | 260 | (191) | - |
| Lindisfarne | 35 | (37) | 0 |
| Big Waters | 3 | (2) | 2 |
| Wallsend Swallow Pond | 1 | (1) | 2 |

Post-breeding gatherings in July included 150 at St Mary's Island on 27th and 50 at Hauxley on 6th. Last records were in October when four were at Newbiggin on 9th, three flew N at the Farne Islands on 15th and one flew N at Annstead on 20th.

The only inland record was two flying W at Whittle Dene on 19th July.

## Arctic Tern, *Sterna paradisaea*

A common breeding summer and passage visitor.

The first arrivals were on 21st April at the Farne Islands, 27th on Coquet Island and 2nd May at Ladyburn Lake.

Breeding data shown below indicates a return to former numbers after the poor weather affected the colonies in 1991:

|  | no. pairs/nests (1991 in brackets) | | no. young fledged where known |
|---|---|---|---|
| Farne Islands | 3,437 | (1,848) | - |
| Coquet Island | 572 | (439) | 309 |
| Mainland Site | 170 | (41) | 75 |
| Lindisfarne | 5 | (0) | 0 |

At the mainland site up to eleven first-summer birds were seen in June and five on 3rd-14th July.

Post-breeding concentrations included 300 at Newbiggin on 30th July and 98 in Beadnell Bay on 4th August. The last sightings were in October with one flying S at Holy Island on 10th and two on the Farne Islands on 15th.

There was an interesting inland record. An adult at Hallington on 1st-2nd August spent most of the time feeding on Sticklebacks, *Gasterosteus aculeatus*, which it seemed to have difficulty turning to get the spines the right way for swallowing.

There were two examples of longevity with birds ringed on the Farne Islands in July 1971 being found dead in 1992 at Co. Down, N. Ireland, and locally on Holy Island.

**Little Tern,** *Sterna albifrons*
An uncommon breeding summer and passage visitor.
    The first returning birds were noted on 25th April on the Farne Islands and on 1st May at Holy Island Causeway. 12 were on Inner Farne beach on 18th May.
    51 pairs bred at a regular mainland site (30 in 1991) raising at least 31 young. On the Lindisfarne Reserve 15 pairs (17 in 1991) attempted to breed but no young were fledged.
    Following the breeding season eight flew S at Newbiggin and six at Seaton Sluice on 23rd July. A late bird, an adult, flew N off Holy Island on 10th October.

**Black Tern,** *Chlidonias niger*
An uncommon passage visitor.
    Spring passage was noted from 14th-26th May at eight localities with four at Caistron on 19th, three at Arcot Pond on 18th and one-two at Hauxley, Big Waters, Castle Island, Druridge Pools, Ladyburn Lake and the Farne Islands.
    August records comprised two juveniles and an adult at Cresswell Pond on 15th-16th, a juvenile at Holywell on 9th and one at Coquet Island.
    A significant influx occurred from 11th-18th September when a total of three flew N and 17 S at Seaton Sluice, eight flew W at the Farne Islands on 11th, seven were near St Mary's Island and one-two at Newbiggin and Big Waters. The last record was one flying N at Seaton Sluice on 23rd.

**Guillemot,** *Uria aalgae*
An abundant breeding species and passage visitor.
    Many birds were around the Farne Islands during the first quarter and in the breeding season 12,912 pairs were counted compared with 14,063 in 1991. This decrease may reflect early breeding by some pairs before the census was made rather than a real decline. A leucistic bird was off the Farne Islands in May. Three pairs also held territory on Dunstanburgh cliffs and three-six birds were at Needles Eye although there was no evidence of breeding. In early July 120 were on the sea at Dunstanburgh.
    Some large movements N of auks occurred in October with 1,500 at St Mary's Island on 4th, 1,200 at Seaton Sluice on 10th and 2,000 per hour at Annstead on 20th. During these movements about 80% were Guillemots.
    One ringed in Grampian as a pullus in 1987 was found dead as an oiling victim at Hauxley in April. Another found on the Farne Islands had been ringed on the Isle of May in 1986.

**Razorbill,** *Alca torda*
A well-represented breeding species and passage visitor.
    A total of 84 pairs bred on the Farne Islands (110 in 1991) and 16 at Needles Eye (12 in 1991). At Dunstanburgh three pairs occupied ledges in April and 16 birds were present in early July.

**Black Guillemot,** *Cepphus grylle*
A rare visitor.
    At the Farne Islands one-three were present on 15th-17th April, one on 2nd May and a bird in full breeding plumage from 6th-13th May.
    Autumn records comprised one-three on 17 dates between 31st August-

26th November at the Farne Islands and one off Holy Island on 16th September and 8th November. One was seen off Stag Rocks on 31st December.

**Little Auk,** *Alle alle*
An uncommon winter visitor, occasionally common after persistent N gales.
There was no repeat of the major movements noted in 1991. The only records in the first two months were one flying N at St Mary's Island on 9th January and one at Craster on 18th February. In March N movement was observed at Seaton Sluice with 13 on 24th and eight on 27th.
In autumn singles flew N at the Farne Islands on 17th September and in October on 10th, 11th and 13th. Two birds were wrecked in December: one was found behind Brownsman Cottage at the Farne Islands on 3rd and another lightly-oiled bird was found at Snab Point on 6th.

**Puffin,** *Fratercula arctica*
An abundant breeding species and passage visitor.
Large numbers had returned to the Farne Islands by 5th April but no population estimate was available for the season as a whole (26,329 pairs in 1989). At Coquet Island, there appeared to be a very large increase in the population to 12,000-16,000 pairs (7,564 in 1991) with 0.95 young fledged per pair, but the high population level will need to be confirmed by a fuller census in 1993. Four pairs (two in 1991) attended burrows at Needles Eye on 30th June, one of which had been excavated since last year.
Some large coastal movements of auks, *Alcidae sp*, were noted. On 4th October 1,500 flew N at St Mary's Island. Specific identification, where possible, indicated that about 15% were this species.
Birds ringed in 1980 in Norway, Sweden and Orkney were found on the Farne Islands in March and April demonstrating interchange between breeding areas.

**Stock Dove,** *Columba oenas*
A well-represented breeding resident.
Several large gatherings occurred in the first quarter with 48 at Walltown in January, 63 at Derwent in March and 82 at the latter locality in April. During the quarter parties of up to 27 were noted at Lindisfarne.
During the breeding season four pairs were at Allenheads and Colt Crag and three at Ninebanks. Pairs were noted at 32 other sites ranging from crags to moorland dales and the coast.
The peak counts in the final quarter were 79 at Holywell and 25 at Portgate on the Roman Wall.

**Woodpigeon,** *Columba palumbus*
A common breeding resident that is abundant in winter.
During the first two months the largest gatherings were 1,000 at Whittle Dene on 12th January and at Belsay on 8th February. Flocks of 100-600 were noted at a further 14 localities.
Spring passage was indicated by four sightings of one-two birds on the Farne Islands from 6th April-22nd May and one bird on Coquet Island on 14th May.
There was one record in autumn on the Farne Islands of a single bird on 15th-16th October.

Few flocks of any significant size were reported in the last three months, the largest being 700 at St Cuthbert's Cave on 1st October.

**Collared Dove,** *Streptopelia decaocto*
A common breeding resident.

The largest gatherings in the first four months included 76 at Low Newton in January, 53 at Hauxley in February while in April parties of 100 were at Throckley and 78 at Benton.

The species ability to breed during winter was shown by a pair nesting in a Christmas tree in Morpeth town centre in December 1991-January 1992.

Coastal movement was shown by singles on the Farne Islands on nine dates in April-May and one on Coquet Island in June.

A flock of 80 at Low Newton in early June were attracted by spilled grain. Some large gatherings occurred during autumn and winter with peaks of 150 at Shiremoor, 79 at Earsdon, 50 at Bamburgh and 27-32 at Hulne Park, Acklington and Corbridge.

**Turtle Dove,** *Streptopelia turtur*
An uncommon passage visitor and casual breeding summer visitor.

Numbers improved slightly from the very poor 1991 season but there have been no breeding records since 1982.

Spring records comprised two at the Long Nanny Burn and one at Coquet Island on 14th May, and singles flying N at Newbiggin on 29th, at Low Newton from 5th-7th June and New Hartley on 9th and flying S at Marden Quarry on 13th.

There was an increase in mid-summer records. At Big Waters one flew E on 27th June and it or another was also seen on 25th July. One was at High Howdon, Wallsend, on 28th June.

In autumn one-two were on Holy Island from 27th September-6th October. Singles were on the Farne Islands on 3rd October and at Hauxley on 10th October.

**Cuckoo,** *Cuculus canorus*
A well-represented breeding summer and passage visitor.

The first arrival was on the early date of 8th April at Felton with the main influx starting from 28th-30th April.

The species seemed to be more plentiful than in recent years with records from 35 widespread localities in May (18 in 1991). The same observation was made concerning the status of the species in Otterburn and Coquetdale by observers who regularly monitor the area.

Juveniles were typically scarce. Singles were at the Long Nanny Burn on 22nd July and at Hauxley in August where one on 2nd was attended by a Meadow Pipit, *Anthus pratensis*. The last record was another juvenile at Holy Island on 3rd October.

**Barn Owl,** *Tyto alba*
An uncommon breeding resident.

Breeding was confirmed at one locality in the SE, where two juveniles fledged, and one site in the N of the county where young were being fed on 22nd July. There were indications of breeding at another site in the SE. None bred in the Border Forests. Birds were noted during the breeding season from April-August at a total of 15 localities (six in 1991).

**BARN OWL – an encouraging increase in records**

The species was seen in about 30 other localities during the year (34 in 1991). Three birds hunting together at Netherton on 10th January was the highest count. One with a broken wing was found dead at Druridge Pools in early March.

**Little Owl,** *Athene noctua*
An uncommon breeding resident.

It was a good year for the species with reports from 28 localities (19 in 1991).

Breeding was confirmed near Wooler and at Brandon and West Chevington. Other sightings during the breeding season were at Bingfield, Carterway Heads, Corbridge, Dinnington, Doddington, High Mickley, Hoppen Kiln, Humshaugh, Hyons Wood, Kirkheaton, Longhorsley, Lucker, Weetwood Hall and Whalton.

There was an encouraging number of autumn records with two at Druridge Pools on 27th August and singles at eight sites in October, including two new ones for the species in the county. Two were near Kirkley Hall in December.

**Tawny Owl,** *Strix aluco*
A well-represented breeding resident.

A very interesting early record was of an adult and four fledged young in Hall Grounds, Wallsend on 16th March. This indicates that the eggs were laid on the exceptionally early date around 16th January.

In the Stocksfield-Dipton study area, it was a poor season with seven pairs attempting to breed (ten in 1991) and raising just four young. Breeding was also noted at Jesmond Dene (four pairs), Beacon Hill, Fontburn (two pairs), Lilburn Burn, Warkworth Lane, and Widdrington. Four birds called in Kyloe Woods in June.

They were also seen or heard in many other widespread localities during the year. From 24th-29th December, in a spell of hard weather, birds were seen hunting in daylight at Stonehaugh, Smalesmouth and Plenmeller Common.

One ringed as a pullus at Slaley in May 1983 was found dead at Rothbury in January.

**Long-eared Owl,** *Asio otus*
An uncommon breeding resident and passage and winter visitor.

A regular roost in the SE held up to two birds from January-March. Daytime sightings included one hunting in the Irthing Valley on 4th January and another in a tree near a busy road at Morpeth on 16th February. A road casualty was found near Haydon Bridge on 5th February.

Breeding was successful at one site in the SE (two-three juveniles) and two sites in the Rothbury and Otterburn areas. Further records in the breeding season comprised single pairs at sites in the SW and in the Colwell area, two sites occupied in the N and single birds at three sites in the SE and two sites in the Rothbury and Otterburn areas. In addition, four-six birds had been at another SE site in April and two birds in the Border Forests in February.

An unusual mid-summer record was one which spent two hours on the Farne Islands on 21st June before departing for the mainland. Autumn immigration was noted from September-November with singles arriving from the E at St Mary's Island and Hartley Point and others on the coast at Hauxley, Holy Island and the Farne Islands. Single owls, probably of this species, arrived from the E at Seaton Sluice and Newbiggin on 10th October.

One ringed as a pullus at Kielder in May 1991 was a road casualty at Barnsley, S Yorkshire, in February.

**Short-eared Owl,** *Asio flammeus*
An uncommon breeding species, well-represented as a passage and winter visitor.

Many were on or near the coast during the first four months with counts in the Woodhorn-Newbiggin area of seven in January, five in February and four remaining in mid-March. Seven were also seen in the Border Forests in January with display being noted from 8th February.

One-two were reported from 17 coastal localities in April, seven in May and two in June.

A mixed breeding season followed with shortages of small mammal prey in some areas. At least six breeding pairs in the Otterburn-Simonsides area was the highest in the district for several years but numbers were down elsewhere. Nine pairs in territory on the SW moors compared with 11 in 1991. Five pairs were in the Irthing Valley and two at Colwell.

The first autumn coastal record was a single at Cresswell on 28th-29th July. Immigration occurred in October with singles arriving from E at Tynemouth, Seaton Sluice, Newbiggin and Holy Island. One-two were also on the Farne Islands on ten dates between September-November.

During October-December birds were settled in wintering areas with up to five at Wallsend Swallow Pond, three at Druridge Pools and Prestwick Carr and two at seven other localities. The only late upland record involved two at Ouston Fell in December.

**Nightjar,** *Caprimulgus europaeus*
An uncommon breeding summer visitor.

The encouraging results of the survey organised by the British Trust for Ornithology to determine the breeding status of this species are given in a paper at the end of this report.

**Swift,** *Apus apus*
A common breeding summer and passage visitor.

The first arrival was at Druridge Bay on 21st April followed by singles at Bywell, Holywell and Corbridge by the end of the month. Numbers had increased by May 15th-27th when 150 were at Big Waters and Holywell and 50-100 at Arcot and Bellingham. A large N movement was noted on the Otterburn Ranges on 10th and 36 flew N between 11th-31st at the Long Nanny Burn. In June 200 were at Walbottle and Holywell and 327 flew N at the Long Nanny Burn during the month.

Counts at colonies included 20 pairs in Blanchland village, 17 at West Wylam and two-eight at six other sites. Young fledged from a nest at Stocksfield on the late date of 6th September.

Significant movement S started on 12th July at Seaton Sluice, Druridge Bay and St Mary's Island. 120 were at Styford and 100 at Alnwick and Big Waters between 17th-19th and 120 flew S at the Farne Islands in five minutes on 25th.

Movement continued in August with 250 flying S at Seaton Sluice on 2nd and 400 at Holywell on 15th. A large drop in numbers was noted from 10th-12th at Longbenton. There were only three records after September 20th, the last one being in Newcastle city centre on 30th.

**Kingfisher,** *Alcedo atthis*
An uncommon resident and rare breeding species.

There were reports from seven localities in January including three in Jesmond Dene.

There was some revival in the breeding population this year with reports from 18 widespread inland localities between March-August (13 in 1991). There was no confirmation of success but two-three birds at Hexham, Powburn and Wylam in August was encouraging.

There were records from 23 localities between September-December (ten in 1991) with two birds at Hallington and Bellingham in October. One at Bakethin from 15th-17th November was seen to catch four Eels, *Anguilla anguilla*, three of which were too large and were released.

**Hoopoe,** *Upupa epops*
A rare passage visitor.

One was at Bamburgh from 2nd-7th April and another at Newbiggin on 30th September.

**Wryneck,** *Jynx torquilla*
A rare passage visitor.
The only record of this increasingly scarce visitor was a single on Holy Island on 27th October.

**Green Woodpecker,** *Picus viridis*
A well-represented breeding resident.
All indications are that the status of the species was unchanged.
'Yaffling' was noted from 8th February. Reports were received from 33 localities between March-July (34 in 1991). Breeding was confirmed at Harehope and Stocksfield.
Outside the breeding season there were records from a further ten localities (eight in 1991), including Holywell on 27th November.

**Great Spotted Woodpecker,** *Dendrocopos major*
A well-represented breeding resident and uncommon passage visitor.
They were again reported from many widespread localities with 'drumming' noted from 8th February. Breeding was confirmed at Hyons Wood (three pairs), the Harthope Valley, Priestclose Wood, Walbottle, Arcot Pond, Horsley Wood and Wheelbirks Farm, Stocksfield.
The only indication of passage was one flying SW over the Farne Islands on 26th October.
In the final two months there were indications of an increasing population with singles in unusual sites at Tepper Moor and Chapel House Estate, Newcastle. Birds fed on garden feeders at seven localities during the year.

**Lesser Spotted Woodpecker,** *Dendrocopos minor*
A rare visitor.
It was a poor year for sightings. The only records were of single females in the N at Ford on 8th February and in the SW at Newbiggin, Hexham on 9th August.

**Skylark,** *Alauda arvensis*
An abundant breeding and passage visitor.
Lindisfarne held the highest numbers at the start of the year with flocks of 100 on 20th-22nd January and 85 on 1st February.
Numbers inland increased during February with singing birds noted at Big Waters on 8th and in upland areas from 21st. In March at Ilderton Dod 40 were present on 14th and 30 moved W on 21st. One-four were noted on the Farne Islands and Coquet Island in April.
Breeding data included 20 singing birds at the Long Nanny Burn in May.
Autumn passage to S-SW commenced in September with movements noted at Ordley on 12th-28th, Warkworth on 13th (25 per hour), the Farne Islands from 18th, Cramlington on 20th and Holy Island on 30th (total 200). Movement continued into October with 30 flying S at Hauxley on 3rd and, on 8th, 80 flying N over the Farne Islands and 15 arriving off the sea at Newbiggin. 126 were on Holy Island on 11th with a further 70 coasting S.
Two large wintering flocks were noted at Lindisfarne in December: 500 on Holy Island on 30th and 150 at Fenham Flats on 29th.

**Shore Lark,** *Eremophila alpestris*
An uncommon winter visitor.
  The only record was a single found at Newbiggin on 25th December which was seen regularly until at least 4th April 1993.

**Sand Martin,** *Riparia riparia*
A common breeding summer and passage visitor.
  The first arrivals were in late March with two at Whittle Dene on 23rd and one at Ordley on 26th. Movement N-NW was noted at Big Waters and Corbridge between 12th-15th April as many more passed through and arrived in breeding areas. By the end of April 250-300 fed at Corbridge and 150 at Big Waters.
  Colony counts included 170 birds at Togstone Links, 60 at Hauxley Reserve, 25 at Long Nanny Burn and 19 at Belsay with smaller groups in several other areas. Five pairs were in cliffs and dunes between Boulmer-Howick.
  Post-breeding gatherings occurred in mid-July with parties of 80 at Whittle Dene and Druridge Pools. Peak concentrations followed between 14th-31st August with 250 at Prudhoe, 200 at Hexham and two flocks of 100 in the Druridge area.
  Most had gone by September with only one-four being seen in four localities. The last record was a single flying N at Newbiggin on 5th October.

**Swallow,** *Hirundo rustica*
A common breeding summer visitor, abundant on passage.
  The first arrivals moved N at Ordley on 10th April and Sipton Burn on 11th. By 6th-8th May 100 had gathered at Arcot Pond and 90 at Druridge Bay Country Park. Strong N movement was noted on 4th at Stag Rocks with 100 per hour. At the Long Nanny Burn a total of 373 flew N from 11th-31st May.
  For the third year in succession, a pair nested on the Farne Islands, raising two broods. Breeding was also confirmed at a shed on moorland in the SW at 360 mts a.s.l.
  Movement S was conspicuous from 28th July-1st August when 100 passed the Long Nanny Burn, steady passage was observed in Budle Bay and 400 were on Holy Island. Large roosts are often found in early autumn and this year 1,000 were at Newton Pool on 3rd August, 600 at Broomlee on 11th, 300 at Newbiggin on 30th July and several hundred at Chevington on 3rd September.
  Large gatherings were still present in September with 300 at Matfen on 5th and 250 at Corbridge on 14th. The last significant movement of the year was 13 flying S at Newbiggin on 1st October. Numbers were then much reduced with 15 at Harbottle on 11th the highest count followed by records of one-two birds from four localities in November, including a very late bird in Budle Bay on 29th which was probably the individual noted again at Bamburgh on 1st December.
  One ringed at Druridge Bay Country Park in September 1990 was controlled at Icklesham, Sussex, in September 1991.

**House Martin,** *Delichon urbica*
A common breeding summer and passage visitor.

The first reports were from Wallsend Swallow Pond where there were two on 7th April and 12 on 8th. The main arrival started on 20th with 25 at Big Waters by 30th and ten-15 at four other localities. Numbers increased during May with 70 at Holywell on 27th and 40-50 at Arcot and Warkworth Lane.

Colony counts included 60 nests at Southfield Green, Cramlington, 35 nests on the Royal Border Bridge, Berwick, seven nests on the Plough Inn, Beal, five-six pairs in Holy Island village and four at Cullernose Point. 21 birds returned to the colony at Kirkhill, Morpeth, and, for the first time for 15 years, a pair nested at Killingworth Lake House.

The largest post-breeding concentrations in early autumn were 200 at Druridge Bay Country Park on 22nd July and 100-120 at Whittle Dene, Prudhoe and Cresswell from 22nd August-5th September. A very large gathering, peaking at 500, was near Close House in dense fog on 26th September and the next day 100 were at Cramlington.

Numbers were much reduced by October with the highest count of 12 at Belsay on 18th. The last report was of two at Mitford on 21st October.

**Tree Pipit,** *Anthus trivialis*
A common breeding summer visitor, well-represented on passage.

The species was scarce in April with just three reports of singles at Slaley on 8th and 22nd and Bellingham on 9th. Coastal passage was noted from 10th-28th May with two on Holy Island and at the Farne Islands and singles at Bamburgh and Prior's Park. One at Arcot Pond on 10th-23rd May was the first recorded at this locality.

From late May-early July displaying and singing birds were noted in 15 typical breeding areas with the highest counts of ten in Thrunton Wood, eight in Dipton Wood, seven in the Harthope Valley and six in Redesdale Forest.

There were more autumn records than usual. On the Farne Islands one-six were noted on 15 dates from 1st August-30th September. One was in the Pegswood-Stobswood area on 18th August and 2nd September. From 26th September-3rd October three were on Holy Island and two at Hauxley and Prior's Park.

**Meadow Pipit,** *Anthus pratensis*
An abundant breeding species and passage visitor, well-represented in winter.

Small parties of overwintering birds were noted in upland areas at Lambley, Broomlee, Blanchland Moor and Knarsdale during the first two months.

A major influx on a broad front was noted from 27th March-5th April with 150 at Bellingham and steady N movements at Ordley, Seaton Sluice, Newbiggin, Stag Rocks and Newton Point. Despite extensive snow cover many were back in breeding areas on the Simonsides by 5th April.

Post-breeding flocks were reported from 7th August on the SW moors. Movement S was first noted on 5th September at Cramlington. This passage intensified on 12th-13th on the coast at Newbiggin, Warkworth, North Shields, Bamburgh, Blyth Park, Druridge and Hauxley and inland at Ordley and Big Waters. From 19th-30th feeding flocks of 100-300 were at Newbiggin, Tynemouth, Kielder Dam and the Irthing Valley.

Few remained in the last three months. Five-17 were at four upland localities in October and November and singles remained at Edmundbyers Common and Yarrow Moor in the freezing weather in late December.

**Rock Pipit,** *Anthus petrosus*
An uncommon breeding resident, usually well-represented on passage.
High counts during February included 45 on Holy Island, 16 at Boulmer and ten at Low Newton and St Mary's Island.
15 pairs bred on the Farne Islands (19 in 1991) and one pair at Coquet Island. Four birds displayed on the coast between Craster-Cullernose Point in May.
Passage was conspicuous in October when 69 were around Holy Island on 10th-11th, 20 at Annstead Beach on 10th, 12 at Hauxley and 11 at Craster.
There were more records than usual of the Scandinavian Rock Pipit, *A.p. littoralis*, which is a rare passage visitor. Two were at Holy Island on 1st February and one at Newton Pool from 29th February-31st March. Also in March, singles were at St Mary's Island on 14th, Holy Island on 15th and Hauxley Reserve on 18th-30th. From 3rd-5th April two-four were on St Mary's Island.

**Water Pipit,** *Anthus spinoletta*
A rare visitor.
The only record was a single at Whittle Dene on 28th March.

**Yellow Wagtail,** *Motacilla flava flavissima*
A well-represented breeding summer and passage visitor.
The first arrival was one flying over Ladyburn Lake on 12th April. A more general arrival commenced on 20th April with one-five birds at ten localities by the end of the month.
Inland seven pairs bred at Whittle Dene, two at Coatenhill and singles at Sparty Lea, Catton, Felton and West Heddon. On or near the coast, single pairs bred at Old Hartley, Buston Barns, Wallsend Swallow Pond and possibly Tilmouth Park, Berwick. Four-six birds were at Cresswell Pond and Druridge Pools in late June.
Gatherings in August included 25 at Whittle Dene, ten at Cresswell Pond and eight at Anick. Numbers declined in September when the highest count was six at Newbiggin on 15th. The last record was a single at St Mary's Island on 24th September.
Blue-headed Wagtails, *M.f. flava*, are rare summer visitors which occasionally breed. One was at Cresswell Pond on 4th May. On 26th July at Hauxley a male accompanied a female of the British race, *M.f. flavissima*, and six juveniles. This is the third consecutive year in which this race has been proved to breed in the county.
Grey-headed Wagtails, *M.f. thunbergi*, are rare summer visitors. Single males were on the Farne Islands on 13th May and at Wallsend Swallow Pond on 31st May. These are the 17th-18th records, the last being in 1990.
Ashy-headed Wagtails, *M,f, cinereocapilla*, are extremely rare summer visitors. A male at Holywell Pond from 26th-30th May and on 22nd June was the second county record, the only other being in 1984.
Sykes Wagtails, *M.f. beema*, are extremely rare visitors. For the first time a male showing characteristics of this race was recorded on the Farne Islands on 28th May.

**Grey Wagtail,** *Motacilla cinerea*
A well-represented breeding species that is uncommon on passage.

The species was noted throughout the year at many inland riverine localities but, unusually, there were no indications of spring passage on the coast.

In the breeding season four pairs were in the Harthope Valley, three on the Wansbeck in the Morpeth-Mitford area and two on the Pont in the Medburn-Ponteland area.

A juvenile in Embleton Bay on 25th July was the first indication of autumn passage. Several migrant families were in forest clearfells in Kielder Forest on 18th August. Coastal movement was noted from 12th September-23rd October with the highest counts of four flying S at Prior's Park on 12th September and three S at Hauxley on 19th.

A number of birds over-wintered in Newcastle city centre with one-four in the St Nicholas Cathedral area and in Grainger Street in December.

**Pied Wagtail,** *Motacilla alba yarrelli*
A common breeding species and passage visitor. Well-represented in winter.

As usual, numbers were low in the first two months with 28 in the Newburn-Wylam area on 10th January being the largest concentration.

Passage was conspicuous from 21st-28th March with 40 on the coast between Hauxley-St Mary's Island, 12 at Low Newton and light N movements at Harwood Forest and Ordley. Many were back in breeding territories in Coquetdale on 29th March.

A good breeding season was indicated in Coquetdale with large numbers of territories occupied and many young seen in June. Two pairs bred on the Farne Islands (two in 1991) and one on Coquet Island.

Post-breeding gatherings in July included 80 at Newbiggin and 40 at Housesteads. The largest gatherings were noted from late August-early October with 140 at Corbridge on 26th September and 50-60 at Longhorsley, Prudhoe, Bellingham and Peth Foot. Small numbers moved S at Blyth Park on 12th September.

Four birds at a site in Newcastle was the highest count in the last two months.

White wagtails, *M.a. alba*, are uncommon passage visitors and extremely rare breeders.

The first record was one at Low Newton on 21st March followed by singles at Cresswell on 25th and Whittle Dene on 31st.

Spring passage was conspicuous from 17th April-11th May. Exceptionally large numbers occurred on the North Shore, Holy Island, with 18 there on 20th April, ten on 21st, six on 2nd May, 38 on 3rd and at least 100 on 9th. Also on the latter date, 12 were at Hauxley.

For the second year in succession a pair bred on the Farne Islands. A pair were also seen on North Shore, Holy Island, on 12th July.

One-two birds were seen at two inland and six coastal localities in the autumn with the last sighting at Holy Island on 27th September.

**Waxwing,** *Bombycilla garrulus*
An uncommon passage and winter visitor, well-represented in irruption years.

Significant numbers remained at the start of the year but flocks were smaller than immediately after the influx in November 1991. From 1st-5th

January reports were received from five localities in the SE with the highest count of 35 at Gosforth Park. On 15th 35-40 flew over Newcastle Town Moor and a flock of 24 was at Ordley on 19th. Eight were in Hexham and six at Amble Braid in early February.

In Newcastle numbers increased from 12th-17th February when 72 were off Barrack Road and 40-45 in the Jesmond area. The return passage in spring peaked in late March-early April with 150 in the Haymarket area on 28th-29th March and 100-110 in the Coach Lane-Red Hall area on 7th April. Two were in Kielder Village on 16th March.

A rapid exodus then occurred from Newcastle with 40-42 remaining at Coach Lane on 9th-10th April and 50 at South Gosforth on 13th. The last record for the spring was one in Amble on 27th April.

There was no repeat of the 1990 and 1991 autumn influxes. The only records in the last two months were three at Morpeth on 25th November, three-four in Gosforth on 10th December and six at Arcot on 20th December.

**Dipper,** *Cinclus cinclus*
A well-represented breeding species.

They were noted in many widespread typical localities. Three sites were occupied in the Felton area in January and four at Hulne Park in March.

Breeding was confirmed at Lilburn Burn, Threestoneburn, Hulne Park (two pairs), Fontburn, Felton, Mitford, Haydon Bridge, Plankey Mill, Warden and Gilderdale. Single pairs were noted at seven other sites from March-August.

Notable concentrations at the end of the year were seven birds in territory on the Usway Burn in October, six birds in the Alwinton area in November, and eight birds on the Breamish from Beanley-Brandon and three pairs on the Lilburn Burn in December.

**Wren,** *Troglodytes troglodytes*
An abundant breeding resident, well-represented on passage.

Spring passage involved one-four almost daily on the Farne Islands from 5th-25th April and one-four on Coquet Island during March-April.

Breeding densities included 45 territories in Hulne Park, 40 in Jesmond Dene and 18 in Hyons Wood and Plessey Woods. Pairs bred successfully on the SW moors at 380 and 430 mts. a.s.l.

Autumn passage lasted from 20th September-4th December on the Farne Islands with a maximum of 12 birds reported. Other indications of passage were eight birds ringed at Hauxley in September and 22 in October and counts of eight on Holy Island on 22nd October and 20 on 8th November.

Birds were again over-wintering in heather moorland at 350 mts. a.s.l. at Darden Lough.

**Dunnock,** *Prunella modularis*
An abundant breeding resident, well-represented on passage.

Spring passage was indicated by one-six birds seen almost daily from 5th-19th April on the Farne Islands and one-three from 31st March-9th April on Coquet Island.

Autumn passage was noted on the Farne Islands with one-three from 10th September-4th December and on the mainland at Hauxley where 14 were ringed in September (maximum of seven on 13th) and seven in

October. In addition, six were at Craster on 13th September, 12 at Bamburgh on 12th October and one at Hadston Point on 28th November.

An example of longevity was provided by an individual ringed as an adult at Kielder in October 1984 and controlled in the same area in December 1991.

**Robin,** *Erithacus rubecula*
An abundant breeding resident and common passage visitor.

Spring passage was reported from 31st March-5th April with 40 at Newbiggin, 25 at Druridge Bay Country Park, 23 at Hauxley, 20 on Coquet Island and 12 at Low Newton and Warkworth Lane. Individuals showing characteristics of the continental form were at Newbiggin and Low Newton during this period. On the Farne Islands one-seven were seen on most dates in April.

Autumn passage was observed from 20th August-30th November on the Farne Islands with 17 present on 30th September, 20 on 4th October and 12 on 8th. At Hauxley 27 were ringed in September (maximum of seven on 12th) and 34 in October. Other counts on the coast were 14-24 at Craster, Holy Island and Bamburgh from 17th-20th September and 12-35 at Craster, Newbiggin, Blyth Park and Holy Island from 30th September-19th October.

**Nightingale,** *Luscinia megarhynchos*
A rare visitor.

The only record was one flushed from a building on the Farne Islands on 14th May.

**Bluethroat,** *Luscinia svecica*
A rare visitor.

All spring records came from the period 18th-25th May. Single males were on the Farne Islands on 18th and 22nd and at Whitley Bay on 23rd. Two males were on the Farne Islands on 23rd-24th. Single females were at Newbiggin on 24th, Tynemouth on 25th and Coquet Island on 24th-25th.

There was one autumn record, a male at St Mary's Island on 22nd September.

**Black Redstart,** *Phoenicurus ochruros*
An uncommon passage visitor and rare winter visitor.

A female over-wintered at Hadston from January-March and another was at St Mary's Island on 11th-18th January.

The first spring migrant was a female on Coquet Island on 30th March. Peak passage was noted from 31st March-8th April with eight at Low Newton, five at Holy Island, four at Bamburgh and Coquet Island, three at St Mary's Island and one-two at seven other coastal sites. From 11th-13th April one-two were at Hauxley Reserve. On the Farne Islands one-two were present on eleven dates from 5th April-22nd May. Late migrants were seen on Coquet Island with a male on 31st May and a female on 12th June.

Autumn passage was less conspicuous. A female was at Tynemouth on 4th October and one-two were at the Farne Islands on three dates from 4th October-14th November.

**Redstart,** *Phoenicurus phoenicurus*
A common breeding summer and passage visitor.

The first arrivals were in April with one on the Farne Islands on 21st followed by single males at Hyons Wood, Bywell and Kidlandlee Dene on 24th-25th. Further passage was noted on the coast from 27th April-23rd May with one-two birds at six localities. Single late females were on the Farne Islands in June on 6th-8th and 10th.

Breeding in nest boxes included three pairs at Nunnykirk (fledging seven young), two at Bellingham (five young) and singles at Beacon Hill (six young), Linnels (eight young) and Fontburn (six young). 27 young were fledged on the Tarset Burn. In natural sites six pairs bred in the Harthope Valley and one-two pairs at Dipton Woods, Prestwick Carr, Coe Crags, Thrunton Wood and Eals Bridge. In old Ash trees *Fraxinus excelsior*, three pairs bred at Bradford House, Belsay, and one at Ordley.

Autumn movement was first observed at Big Waters on 20th July. In August two were at Hauxley on 1st and one on 7th and singles at St Mary's Island and Newbiggin on 23rd-26th. Passage increased during September-early October with up to eight on Holy Island, six on the Farne Islands, three at Newbiggin and four individuals ringed at Hauxley. The last sighting was one on Holy Island on 10th October.

**Whinchat,** *Saxicola rubetra*
A common breeding summer and passage visitor.

A male at Hauxley on 25th April was the first arrival followed by another at Holy Island on 29th. Coastal passage in May was indicated by 12 at Druridge Pools on 14th and one-three at a further 15 sites.

As confirmed by atlas work, this species breeds most commonly in the N. Ten family parties (total of 50-60 birds) were seen on Quarryhouse-Bewick Moor, about seven at Clennel Hill, Alwinton, five pairs in the Harthope Valley and two-three in the Threstoneburn-Ilderton Dod area.

Autumn passage was first noted on 8th August at the Farne Islands and 19th at Coquet Island. On the mainland coast the highest count in August was 11 at Newbiggin on 30th. Passage peaked from 12th-14th September with 20 at Druridge Plantation, 16 at Hadston Carrs, ten at St.Mary's Island and five-nine at seven other localities. The last sightings were singles on 4th October at Holy Island and on 6th at Newbiggin.

**Stonechat,** *Saxicola torquata*
A rare breeding resident and uncommon passage visitor.

During the first quarter one-three birds were in eight coastal localities (12 in 1991) with just two inland records (six in 1991): a pair at the Simonsides on 12th January and a male at Derwent on 4th March.

Breeding records again came mainly from the favoured Druridge Bay area where two fledged broods were noted at three localities, Druridge Pools, Chevington Burn and Druridge Bay Country Park. Elsewhere on the coast a pair was at Bamburgh in early April and inland a pair bred successfully in Kielder Forest.

In September four birds were at Bamburgh on 19th-20th and one on the Farne Islands on 26th-27th.

In the last three months one-four birds were in ten coastal localities (eight in 1991) and inland a pair was in Kielder Forest on 9th December.

**Wheatear,** *Oenanthe oenanthe*
A common breeding summer and passage visitor.

The first arrival was a male at Newbiggin on 14th-15th March followed on 18th by two at Hauxley and St Mary's Island. The first inland record was a male near Slaley on 31st March.

Spring passage was heavy from 23rd April-15th May with counts on Holy Island of 40 on 23rd April, 31 on 2nd May, 52 on 3rd, 89 on 8th, 176 on 9th and 66 on 13th. Elsewhere on the coast 70 were between Low Newton-Long Nanny Burn on 9th May, 40 at Newbiggin on 7th and 30 at Coquet Island on 4th. Almost daily counts of one-16 birds were made on the Farne Islands from 5th April-28th May.

Breeding was confirmed at a number of inland localities and on the coast at Newbiggin, Longhoughton Steel, Long Nanny Burn, Old Hartley and N. Blyth.

Autumn movement was most conspicuous from 31st August-24th September when 24 were at Cresswell, 15 on the Farne Islands, 14 at Holy Island, seven at Newbiggin and one-five at a further 14 coastal localities. The last inland record was a single at Butterburn on 16th September. Final records were in October with singles at Beadnell and Newbiggin on 11th-12th, the Farne Islands on 16th and Holy Island on 17th.

Some reports mentioned that many individuals observed in the heavy passage from 2nd-15th May showed characteristics of the Greenland race, *O.o. leucorrhoa.*

**Ring Ousel,** *Turdus torquatus*
A well-represented breeding summer visitor, uncommon on passage.

The first arrivals were on the late date of 8th April at Sipton Burn. Another was at Langleeford on 11th. On the coast passage was noted at the Farne Islands, Craster, Holywell Pond and Swallow Pond between 21st-28th. During May, one was on Bamburgh Golf Course on 12th.

Breeding was confirmed at several typical localities including three pairs at Gilderdale in June and four-five birds at Carey Burn and Linhope. Two were still well inland on high moorland near Coalcleugh in early October.

Autumn passage commenced in September with a bird arriving from E with Song Thrush, *T. philomelos,* at Bakethin on 20th. On the 30th there was an influx on the coast involving five at Dunstanburgh, four on the Farne Islands, three at Newbiggin and two on Holy Island. In October two were at Newbiggin and singles on the Farne Islands and at the unusual location of Westerhope from 1st-4th. Further singles were at Roscastle on 16th and Hauxley on 23rd. Three at Hauxley on 15th November were the last recorded.

**Blackbird,** *Turdus merula*
An abundant breeding resident, passage and winter visitor.

During March, an increase in numbers was noted inland at Ilderton Dod and Threestoneburn between 26th-27th, some of which were seen to move E. Coastal passage was indicated from 2nd-4th April when 30-40 were at Newbiggin and 12 at Holywell Dene.

Autumn migration commenced during September with ten at Horton on 20th and 100 at Newbiggin on 30th. Passage in October was unexceptional with the first influx between 3rd-7th involving 100 at Brier Dene and 30-40 at

Blyth Park, Newbiggin and Whitley Bay and a second influx from 26th-31st involving up to 40 at Newbiggin. Immigration continued into November with a fall on 8th-9th of 1,350 on the Farne Islands, 60 at Holy Island and many coming off the sea between Bamburgh-Seahouses. Another small fall on 15th resulted in 31 birds being ringed at Hauxley.

A partial albino male was at Swallow Pond on 17th March.

The Scandinavian origins of many of our autumn visitors is shown by the fate of five birds ringed at Hauxley in October between 1988-91. In 1992, singles were found dead in Goteborg, Sweden in February, Jylland, Denmark in March, Ostfold, Norway in May and one was controlled at Vest-Agder, Norway in November. The fifth individual was found dead at Indrearna, Norway in September 1991.

**Fieldfare,** *Turdus pilaris*
An abundant passage and winter visitor that has bred.

The largest flocks during January were 400 at Sharperton and Kirkharle. In February 400 were at Harwood and 300 at Allenbanks.

Numbers increased during March when 'thousands' moved NW on 27th from Redesdale towards Carter Bar, 1,000 were in the Otterburn-West Woodburn area, 600-700 at Wark Forest and 100-500 at another 14 localities. They remained conspicuous during April with pre-emigration flocks of 500 at Chatton, 400 at Harwood and Netherton, 380 in Hexhamshire and up to 300 at ten other localities. By May numbers had dropped sharply, one at Newbiggin on 8th being the last.

An interesting summer record was of two birds at Whitley Bay Cemetery on 10th July.

A single on the Farne Islands on 9th August was the first returning bird. 20 were on Alnham Moor on 10th September. Very few were noted early in October but numbers increased from 21st when 200 flew W over Mitford and a marked influx occurred on Holy Island from 25th-26th. 550 at Ayle on 1st November included fresh arrivals spiralling out of the sky from the NE and 500 were on Dunterley Fell, Bellingham, on 3rd. The largest late flock was 1,000 at Kirkharle on 29th November when 400 were at Whittle Dene. 200 roosted at Cupola Bridge in December.

**Song Thrush,** *Turdus philomelos*
A common breeding species and passage visitor.

Two were in full song at Exhibition Park, Newcastle, on 8th January. During February-March, birds were noted returning to several inland breeding areas.

Spring passage was indicated by 20 at Newbiggin on 2nd April, one-three at the Farne Islands on nine dates and one at Beacon Point on 24th.

The first signs of autumn migration were two on the Farne Islands on 19th September, four arriving from the E at Bakethin on 20th and one at Newbiggin on 22nd. Numbers then slowly increased until a major influx occurred on 30th with 1,000 moving S from Holy Island and 300 at Newbiggin. This influx continued into early October with 100 at Newbiggin, up to 40 on the Farne Islands and 12-30 at Bamburgh and Blyth Park. During December, a bird was in song at Cochrane Park, Newcastle.

Rapid movement of Scandinavian migrants was shown by a bird found dead on the Farne Islands in October, four days after being ringed in Norway. Onward movement to Iberian wintering areas was shown by an individual ringed at Hauxley in October 1991 being shot on the Algarve, Portugal in January.

**Redwing,** *Turdus iliacus*
An abundant passage and winter visitor that has bred.

During the first two months, numbers were unexceptional, the largest flocks being 40 near Whittle Dene in January and 67 at Jesmond, 56 at Ordley and 40 at Morpeth in February. March saw a flock of 325 at Ordley on 14th decline to 60 by 28th as passage began.

Emigration in April was indicated by one-six at Newbiggin and Hauxley between 2nd-3rd and birds heard flying out to sea at Blyth on 4th and Newbiggin on 9th. The last record was a flock of 50 at Jesmond Dene on 10th flying E.

The first autumn arrivals were five at Hauxley on 19th September with one-30 at six other localities on or near the coast by the end of the month and one inland at Hexham on 22nd. Arrivals continued throughout October, with peak numbers including a few hundred moving W over the Border Mires on 7th, 170 at Allen Banks on 10th, 180 at Stamfordham on 25th and 100 at Belsay on 31st. During November, peak numbers occurred between 1st-8th with 150 at Corbridge, 135 at Ayle, 125 at Hyons Wood and 120 at Slaley Hall.

Numbers increased again during December with 350 at Belsay Hall, 200-250 at Brunton Banks, 200 at Corbridge and 20-150 at another nine widespread localities.

**Mistle Thrush,** *Turdus viscivorus*
A common resident breeding species, uncommon on passage.

11 at Horton on 5th January and in the Blanchland area on 4th February were the largest early gatherings.

In May and June breeding was confirmed at Gosforth Park, Hareshaw Lynn, Holywell, Horton, Long Nanny Burn, Matfen, Threestoneburn, Wideopen and was strongly suspected at Newcastle Central Station.

Some large post-breeding gatherings followed with 54 at Whittle Dene, 39 at Thrunton Woods and 35 at Beanley Wood in July. In September 63 were at East Bolton and coastal passage was indicated by 12 arriving high from SE at Bamburgh and gatherings of 14 at Howick and nine at Craster. During October one flew over the Farne Islands, 30 were at Newbiggin and 13 at Bamburgh.

**Grasshopper Warbler,** *Locustella naevia*
An uncommon breeding summer and passage visitor.

The first arrivals were singles 'reeling' at Arcot Pond on 21st April and near Big Waters on 24th. Between 26th-30th one-three were in a further six localities.

During May five were 'reeling' in the Ponteland-Medburn area between 1st-8th with further records of three at Big Waters on 1st, Druridge Bay Country Park on 3rd and Druridge Pools on 21st. One-two were settled in another 14 widely scattered localities.

In July three were still 'reeling' at Prestwick Carr on 14th with a further one-two at another nine widespread localities. Breeding was confirmed at Gosforth Park where two juveniles were ringed.

The last records were in September with singles on the Farne Islands on 2nd and Cocklawburn on 29th.

**Savi's Warbler,** *Locustella lusciniodes*
An extremely rare visitor.
One was at Coquet Island from 24th-25th May. This is the second county record, the first being at Beal in June 1977.

**Sedge Warbler,** *Acrocephalus schoenobaenus*
A common breeding summer and passage visitor.
Two on the Farne Islands on the very early date of 13th April were the first arrivals. From 25th-30th one-four were singing at five traditional breeding areas. During May and June they were singing in many localities with 15 in the Medburn area, 11 at Alnmouth Links, nine at Hauxley, eight at Big Waters and six at Druridge Pools. In July successful breeding was noted in several localities.
By late summer birds were returning S with 33 at Bamburgh from 26th-30th July and three at Druridge Bay Country Park on 8th August. In September three were at Newton Pool on 19th and singles were at five mainly coastal localities. The last was at Big Waters on 27th.
A fascinating record involves an individual ringed at Gosforth Park in June 1991. It was controlled on return passage near Agadir, Morocco, on 24th April and was recognised from colour rings back at Gosforth Park by 14th May. A bird ringed in Kent in August 1990 was controlled at Big Waters in May. Another bird ringed at St Philbert de Gr. Lieu, France, in September 1989 was controlled at Druridge Bay Country Park in May 1990.

**Marsh Warbler,** *Acrocephalus palustris*
An extremely rare visitor.
It was a superb year for this species. Two were on Holy Island and one at Newbiggin from 7th-8th June with another single at Coquet Island on 2nd.
In September singles were at St.Mary's Island from 22nd-23rd and at Newbiggin on 30th. These are the fourth-ninth records for the county, the last being at Hauxley in September 1980.
A single unstreaked *Acrocephalus* warbler on the Farne Islands on 29th October may have been this species.

**Reed Warbler,** *Acrocephalus scirpaceus*
A rare breeding species, uncommon on passage.
The first was at the favoured locality of Gosforth Park on 4th May where two-three were present by 25th. Singles were at Linnshiels Lake on 13th, Druridge Bay on 14th, Coquet Island on 21st, the Farne Islands from 28th-29th and Holy Island on 31st.
Between June-August, up to six birds were singing at Gosforth Park, two-three at Big Waters, two at Holywell Pond and Weetslade Pond and one at Annitsford Pond. Breeding was confirmed at Gosforth Park.
Autumn migration was noted in September with singles at St.Mary's Island from 23rd-24th and Hauxley on 26th-27th. The last was one at Craster on 4th October.
Single unstreaked *Acrocephalus* warblers, probably this species, were on the Farne Islands on seven dates between 17th-30th September.

**Icterine Warbler,** *Hippolais icterina*
A rare visitor.
It was an exceptional year for this species. The first arrival was on the

Farne Islands on 28th May. A spring fall was noted in June with two on Holy Island from 7th-8th, one at Marden Quarry on 7th and different individuals on the Farne Islands from 8th-11th and on the 9th.

Autumn passage started on 30th August when singles were at St.Mary's Island and Tynemouth. In September singles were at the Farne Islands on 5th and 13th and one was ringed at Hauxley on 13th.

### Subalpine Warbler, *Sylvia cantillans*
A rare visitor.

A male was at Newbiggin on 13th May. This is the 11th record for the county and has been accepted by the British Bird Rarities Committee.

### Barred Warbler, *Sylvia nisoria*
A rare visitor.

Singles on Coquet Island and the Farne Islands on 9th August were the earliest autumn arrivals so far recorded. In September single juveniles were noted at Newbiggin on 18th and Newton Point on 27th. One was at Craster on 13th October.

### Lesser Whitethroat, *Sylvia curruca*
An uncommon breeding summer and passage visitor.

The first arrivals were singles on Holy Island on 25th April and Arcot Pond on 30th. They were present in good numbers during May with birds in at least 18 widespread localities, including four holding territory at Arcot Pond and Felton Lane.

During the summer they continued to be well-reported and breeding was confirmed at Ancroft, Arcot Pond, Felton Lane, Holywell Pond, Marden Quarry and Stamfordham.

Return passage started in August with one-two noted at several coastal localities. Two were on Holy Island on 17th October.

The last record was of one on Holy Island on 24th October which showed characteristics of the Siberian race, *S.c. blythi*. This is the 12th record for this race, the last being in 1988.

### Whitethroat, *Sylvia communis*
A common breeding summer and passage visitor.

Singles at Pegswood on 23rd April, the Farne Islands on 25th and Backworth on 28th were the first arrivals. By May they were noted as being more numerous than in 1991 with birds noted in at least 38 widespread localities, including eight at Felton and seven at Corbridge.

Successful breeding was noted in several localities from June with three broods at Felton Lane and two broods at Arcot Pond and Alcan ash lagoons, Newbiggin.

Autumn passage in September involved one-three at nine coastal localities and the Farne Islands. The last records were on 30th when there were three at Newbiggin and one on the Farne Islands.

### Garden Warbler, *Sylvia borin*
A common breeding summer and passage visitor, extremely rare in winter.

One at Morpeth on 1st February was the third winter record for the county, the last being in 1976 at the same locality.

In April the first spring arrivals were at Druridge Pools on 9th, Swallow

Pond on 26th and Hexham on 26th-27th. Passage continued into May with an influx observed from 14th. By 31st birds had been noted at 24 widespread localities including seven at Lambley, five at Beacon Hill and Holy Island and four at Stocksfield. Four were involved in a fall on Holy Island on 8th June.

Successful breeding was confirmed at Allenbanks and Bellingham. Autumn migration on the coast was noted from 20th-30th September with individuals at six locations. During October singles were at Blyth Park on 3rd, Holy Island on 4th, near Seaton Sluice on 7th and Hauxley on 31st. The last record was one at Hauxley from 15th-16th November.

**Blackcap,** *Sylvia atricapilla*
A common breeding summer and passage visitor, uncommon in winter.

Singles were seen in seven widespread localities, mainly gardens, during January-March.

Spring passage commenced in April with an influx from 16th. By the end of the month peak numbers were 31 males singing in Jesmond Dene, nine between Allenbanks-Plankey Mill, eight at Mitford and one-six at a further nine localities. Three birds were noted in the spring fall on Holy Island from 8th-9th June.

Coastal passage was conspicuous during September with one-15 seen at 13 localities and 23 birds ringed at Hauxley during the month. This passage continued into October with one-13 at ten localities and monthly ringing totals of 21 at Hauxley and six at Bamburgh.

Exceptionally high numbers were seen in November-December with three birds at Jesmond Dene, two at Newbiggin and Spindlestone and singles at 14 other sites, mainly suburban gardens. Four were ringed at Hauxley on 15th November.

Passage of Scandinavian birds was shown by an individual ringed at Revtangen, Norway, in October 1990 being controlled at Hauxley three days later.

**Arctic Warbler,** *Phylloscopus borealis*
A rare visitor.

One was on the Farne Islands on 29th September. This is the 12th record for the county, the last being in 1981. It has been accepted by the British Bird Rarities Committee.

**Yellow-browed Warbler,** *Phylloscopus inornatus*
A rare autumn visitor.

The first arrivals of the autumn were singles at the Farne Islands on 25th September and on Holy Island between 26th-30th.

In October, the traditional peak month for this species, up to two were on Holy Island from 4th-11th with further singles at Hauxley on 2nd, 4th and 6th, Blyth Park on 3rd, Whitley Bay Cemetery on 5th and Holywell Dene on 7th. One was ringed at Hauxley on 13th October.

**Bonelli's Warbler,** *Phylloscopus bonelli*
An extremely rare visitor.

One at Prior's Park, Tynemouth, on 6th October was the second record for the county, the only other being at Hauxley in November 1967. It is still under consideration by the British Bird Rarities Committee.

**Wood Warbler,** *Phylloscopus sibilatrix*
A well-represented breeding summer visitor, rare on passage on the coast.

The first arrivals were five between Allenbanks-Plankey Mill and one at Dipton Wood on 25th April with a further single at Beacon Hill on 30th. During May eight were singing at Beacon Hill on 14th, five at Cupola Bridge on 17th and four at Hareshaw Linn, Holystone Oaks and Stocksfield. Passage birds were noted at the Farne Islands, Newbiggin, North Shields and Stobswood.

One was still singing at West Woodburn on 2nd August. Coastal passage in autumn involved singles at Holy Island on 22nd September, Whitley Bay Cemetery on 28th and Ridley Park, Blyth, on 4th October.

**Chiffchaff,** *Phylloscopus collybita*
A well-represented breeding summer and passage visitor, uncommon in winter.

In January one visited a Holy Island garden on 1st and 5th while another narrowly escaped being killed by a Merlin, *Falco columbarius*, at Newton Pool on 25th. In February single birds were at Big Waters on 8th and again at Newton Pool on 20th.

The first singing bird was at Sandhoe on 17th March with others heard at a further ten widespread localities by the end of the month. They continued to be well-reported throughout the spring with one-four birds at 19 widespread localities in May.

Return passage commenced from mid-August although there were several reports of singing birds up to October. The heaviest influx occurred from 24th-25th September when seven were on Holy Island and four at Felton Lane. Two were still present in November at Newbiggin with singles at Big Waters, Hauxley, Holywell Pond and Swallow Pond. During December singles were at Newbiggin on 5th and 16th, Newton-by-the-Sea on 16th, Gosforth on 24th and 31st and Tynemouth on 25th.

One showing characteristics of the northern race, *P.c. abientinus*, was at Newbiggin on 28th September. Further records of this race were two birds at Hauxley on 6th October and singles at Newbiggin on 29th October and 11th November.

Records of the Siberian race, *P.c. tristis*, in October comprised singles at the Farne Islands on 5th, Newbiggin on 13th and 31st and Holywell Dene from 17th-18th.

**Willow Warbler,** *Phylloscopus trochilus*
An abundant breeding summer visitor, common on passage.

The first arrival was in a Seahouses garden on 6th April with the main influx from 21st. By the end of the month they were widespread with reports from at least 27 localities. Peak numbers involved 35 at Druridge Bay Country Park, 20 at Thrunton Woods, 19 at Hyons Wood and 11 at Wallsend Swallow Pond.

Counts of singing birds during May-June included 24 at Prudhoe Country Park, 15 at Big Waters and Framlington Gate, 14 at Bamburgh and 13 at Wallsend Swallow Pond and Arcot Pond.

Autumn migration commenced in August with small numbers at Blyth Park, Druridge Bay Country Park and singles in gardens at Benton and Walbottle. One-five were noted at ten mainly coastal localities in September. Late passage in October produced singles at Blyth Park, the Farne

Islands, Holy Island, Holywell Dene and Newbiggin. The last record was one at Beadnell on 18th.

One showing characteristics of the eastern race, *P.t. acredula*, was on Holy Island on 30th September.

**Goldcrest,** *Regulus regulus*
An abundant breeding resident, common on passage.

Early in the year they were prominent in many localities, including Harwood Forest where many were singing in February.

Spring passage along the coast started in March with two-15 at four localities. A fall was noted at many coastal localities between 1st-5th April involving 100 at Newbiggin, large numbers at Bamburgh, 23 at Warkworth Lane and two-20 at a further nine localities.

Autumn passage started in late September with small numbers noted on the coast including ten at Prior's Park and three-four at Craster. From 3rd-8th October the main influx occurred with 200-250 on the Farne Islands, several hundred at Newbiggin, 100 at Holy Island, 50 at Ridley Park, Blyth, 45 at Hauxley and one-35 at a further 11 locations.

By the end of the year they were observed as being more plentiful than in late 1991 at Harwood Forest.

A belated recovery involved one ringed at Tellmark, Norway, in September 1990 being controlled at Hauxley 26 days later.

**Firecrest,** *Regulus ignicapillus*
A rare visitor.

A single at Marden Quarry on 6th April was the only record.

**Spotted Flycatcher,** *Muscicapa striata*
A common breeding summer visitor, well-represented on passage.

Two at Bradford House, Belsay, on 13th May were the first arrivals. Later in May birds were noted from 19 widely scattered localities, including up to ten at Beacon Hill on 19th.

Breeding was confirmed at 19 sites although predation was noted at Matfen and Fontburn Reservoir. After the predation of their first clutches, eight pairs at Matfen laid 52 eggs and fledged 31 young. Five broods were at Redesmouth in August.

Autumn migration commenced on 24th August at the Farne Islands. Subsequent records included two at Blyth Park on 30th and one at Hauxley on 31st. In September birds were present on Holy Island from 12th-30th with a maximum of six on 23rd. Four were at Hauxley on 13th where the last sighting was on 27th.

**Red-breasted Flycatcher,** *Ficedula parva*
A rare visitor.

In September single immatures were on the Farne Islands from 18th-19th and from 28th-30th.

**Pied Flycatcher,** *Ficedula hypoleuca*
A well-represented breeding summer visitor and passage visitor.

Three singing males at Dipton Wood on 25th April were the first arrivals. By the 30th three were also at Beacon Hill and two at Allen Banks and Linnels. During May birds were noted in many suitable breeding sites and passage continued with singles at Howick Dene on 9th, Coquet Island from 19th-23rd and Prior's Park on 25th and 30th.

Details from nest-box studies included the following:

| Locality | No. of pairs | No. of eggs | No. fledged young |
|---|---|---|---|
| Dipton Wood | 37 | 246 | 220 |
| Kingswood Burn | 25 | 143 | 102 |
| Linnels | 19 | 133 | 86 |
| March Burn | 15 | 100 | 91 |
| Letah Wood | 8 | 53 | 35 |
| Nunnykirk | 7 | 51 | 49 |
| Dye House | 7 | 51 | 39 |
| Brinkburn | 7 | 50 | 33 |
| Wallington Hall | 7 | 46 | 26 |
| Beacon Hill | 4 | 28 | 20 |
| Fontburn Reservoir | 3 | 16 | 13 |
| Angerton | 2 | 15 | 14 |
| Hartburn | 2 | 15 | 8 |
| Belsay | 1 | – | 7 |

Autumn passage involved birds on Holy Island throughout September with a maximum of ten on 30th, at least five at Prior's Park, two at Craster and sightings at another six coastal localities. One-three present on Holy Island from 4th-7th October were the final records for the year.

There were two interesting recoveries from the ringing study in Kinsgwood Burn. A juvenile ringed in June 1990 was found freshly dead in December 1991 at the Ivory Coast, a movement of 5,257 km. Another juvenile ringed in June 1991 was found dead in April in Morocco, a movement of 2,322 km. The two individuals concerned were reared in the same nest-box in consecutive years by the same mother.

Other recoveries included a pullus ringed at Nunnykirk in June 1991 being controlled as a breeding male at Greenhaugh in June. A pullus ringed at Brinkburn in June, was killed by a cat at Pontefract, West Yorkshire, in August and a pullus ringed at Nunnykirk in June 1990 was controlled as a male in a breeding area at Nannerch, Wales, in September.

**Long-tailed Tit,** *Aegithalos caudatus*
A common breeding resident.

They were noted in many localities during the year. The largest flocks during the first quarter were 45 at Whitley Chapel on 1st February and 20 at Lynemouth Dene on 7th March.

Nest building was observed at Letah Wood on 20th April and successful breeding was reported from 13 widely-scattered localities.

Post-breeding flocks included 26 at Kirkhill, Morpeth, in August and 24 at Felton Lane in September. Noticeable movements occurred throughout the county in October, the most unusual involving ten in a Holy Island garden on 24th.

63 at Belsay Hall on 5th December was the largest of many flocks at the end of the year.

**Marsh Tit,** *Parus palustris*
A well-represented breeding resident.
Records were received of one-five birds from at least 30 widely-scattered localities, mainly in the SE and SW, during the year. Breeding was confirmed at Guyzance, Felton Lane and Howick Dene.

Details from nest-box studies included:

| Locality | No. of pairs | No. of eggs | No. fledged young |
|---|---|---|---|
| Fontburn Reservoir | 1 | 9 | 9 |
| Nunnykirk | 1 | 8 | 8 |

**Willow Tit,** *Parus montanus*
A well-represented breeding resident.
They were reported in small numbers from at least 32 widespread localities during the year, the vast majority from the SE and SW. Breeding was confirmed at Big Waters, Farnley, Felton Lane, Guyzance and Hareshaw Linn.
Unusual records involved singles at Brock Burn, Lindisfarne, in May and at Bamburgh in July.

**Coal Tit,** *Parus ater*
A common breeding resident.
Many were in song in Harwood Forest, Hulne Park and Wallington Hall by February. 30 in the Kidland Forest on 1st March was the largest flock in the first quarter.
During the breeding season the species was noted as abundant throughout Hulne Park where many family parties were noted. Two pairs used nest-boxes at Threestoneburn and single pairs raised nine young each at March Burn and Spindlestone.
Coastal movements during the autumn were light. In September three were at Newbiggin from 12th-16th, two at Tynemouth on 24th and one at Bamburgh on 20th. In October three were at Bamburgh on 24th.
30 at Threestoneburn and 20 at Wallsend Swallow Pond in November and near Blagdon in December were the largest late flocks.

**Blue Tit,** *Parus caeruleus*
A common breeding resident.
The largest gatherings in the first quarter involved 40 at Belsay Hall in February and 30 in a Matfen garden in March. A total of 40 birds were ringed at Matfen in January.
32 pairs were occupying natural nest sites in Hyons Wood by late April and 14 pairs bred successfully at Priestclose Wood, Prudhoe.

Details from nest-box studies included:

| Locality | No. of pairs | No. of eggs | No. fledged young |
|---|---|---|---|
| Dipton Wood | 36 | 354 | 276 |
| March Burn | 20 | 197 | 144 |
| Linnels | 18 | 165 | 125 |
| Nunnykirk | 16 | 127 | 110 |
| Dye House | 12 | 100 | 83 |
| Fontburn Reservoir | 9 | 85 | 75 |
| Beacon Hill | 8 | 78 | 61 |
| Mitford | 7 | 68 | 57 |
| Kirkley Hall | 7 | – | 46 |
| Letah Wood | 6 | 59 | 45 |
| Spindlestone | 6 | 54 | 52 |
| Wallington Hall | 6 | 49 | 30 |
| Big Waters | 4 | 39 | 37 |
| Hartburn | 1 | 10 | 8 |

The largest parties in the final quarter were 45 at Hyons Wood in November and 40 at Belsay Hall in December. 79 were ringed at Sandhoe in November-December and 133 at Matfen in December.

A five-year old bird, ringed at Hauxley in July 1987, was re-trapped at the same site in October. One ringed at Matfen in December 1991 was controlled at Bolam in March. Another, ringed at Lesbury in September, was controlled at Threestoneburn in October.

**Great Tit,** *Parus major*
A common breeding resident.

100 at Belsay Hall on 8th February was the largest party noted in the first quarter.

Details from nest-box study areas included:

| Locality | No. of pairs | No. of eggs | No. fledged young |
|---|---|---|---|
| Wallington Hall | 17 | 123 | 86 |
| Nunnykirk | 14 | 102 | 95 |
| Tarset Burn | – | – | 74 |
| Linnels | 11 | 74 | 64 |
| March Burn | 9 | 72 | 51 |
| Kirkley Hall | 6 | – | 50 |
| Fontburn Reservoir | 5 | 47 | 30 |
| Mitford | 5 | 38 | 16 |
| Big Waters | 4 | 37 | 35 |
| Angerton | 4 | 35 | 35 |
| Dye House | 4 | 27 | 20 |
| Dipton Wood | 4 | 21 | 13 |
| Spindlestone | 3 | 24 | 23 |
| Letah Wood | 1 | 9 | 7 |
| Hartburn | 1 | 7 | 7 |
| Beacon Hill | 1 | 6 | 6 |

A grey-white individual was trapped at Jesmond Dene in November. 46 were ringed at Sandhoe in November-December.

**RED-BACKED SHRIKE – more passed through than in recent years**

**Nuthatch,** *Sitta europaea*
An uncommon breeding resident.
They were seen in 43 widespread localities during the year, including two new sites in the N at Bamburgh and Barrack Wood, Spindlestone.
Breeding was noted at Bywell (family party of eight), Cupola Bridge, Dipton Burn, Hawkhope Burn, near Falstone, Prospect Hill and Wallington Hall where three pairs laid 18 eggs and fledged 11 young.
One trapped at Hexham in March had been ringed as a nestling at Dye House, a movement of 5.6 km.

**Treecreeper,** *Certhia familiaris*
A common breeding resident.
They were noted in many woodland localities. Breeding was confirmed at Beacon Hill and Linnels.
One on the Farne Islands on 10th August was the first indication of autumn movement. Two were at Hauxley from 12th-13th September. Coastal passage continued in October with singles at Bamburgh, Foxton, Holy Island, Lesbury and Seaton Sluice.

**Golden Oriole,** *Oriolus oriolus*
A rare visitor.
A good spring passage was recorded. In May records comprised an immature male on the Farne Islands from 22nd-26th, which was later found dead at the same locality, and single females on Holy Island from 24th-26th and at Arcot Pond on 24th. Sightings continued into June with a female or immature male on Holy Island from 8th-9th.

**Red-backed Shrike,** *Lanius collurio*
An uncommon passage visitor.
It was an excellent spring for this species. In May a female was at St Mary's Island on 24th and single males were at Holy Island on 25th and Chathill from 26th-27th.
The main influx occurred from 5th-10th June with a peak of five females and two males on Holy Island on 8th. Other sightings involved single males at Coquet Island and the Farne Islands and single females at High Newton and two sites on the Farne Islands.

**Great Grey Shrike,** *Lanius excubitor*
An uncommon winter and passage visitor.
One in Kielder Forest on 15th March was the only early record.
Autumn records involved one at Belsay on 10th October and another noted during a snowstorm at Plashetts Burn on 19th November. During December singles were in Falstone Forest on 11th and at Hagg Bank, near Wylam, from 29th-31st.

**Woodchat Shrike,** *Lanius senator*
A rare visitor.
A male was in Wark Forest from 26th-31st May. This is the 13th record for the county, the last being in June 1977 at Holywell Pond.

**Jay,** *Garrulus glandarius*
A well-represented breeding resident.
They were reported at 21 widespread localities during the year. The largest counts were up to ten seen regularly in Gosforth Park during the first four months of the year and eight in Thrunton woods on 6th November.
Birds were reported feeding on peanuts in gardens in Kielder and Hexham.

**Magpie,** *Pica pica*
A common breeding species.
Roosts of over 25 birds were recorded at Wallsend Swallow Pond, Red Row and Linton in the spring. At the end of the year the largest gathering was 40+ at Wallsend Swallow Pond on 11th November.
Urban and suburban breeding was well reported and a pair nested on Holy Island for a second year.

**Jackdaw,** *Corvus monedula*
A common breeding resident.
Flocks of 240 at Derwent Reservoir on 11th January and 200+ at Otterburn were the largest flocks in the early part of the year.
Spring migrants were noted on the Farne Islands on 11th and 19th April.
Autumn migrants were on the Farne Islands on 21st and 24th September. In the latter part of the year the largest roost was of 1,500 birds at Housesteads on 24th August. 1,200 at Lower Heddon on 28th November was the only other flock of over 1,000 birds noted.

**Rook,** *Corvus frugilegus*
An abundant breeding resident.
The largest count in the spring was of 1,000 at Longbenton on 24th January.
Colony counts during the breeding season suggest a slight decline in numbers since the previous year.

|  | 1992 | 1991 |
| --- | --- | --- |
| Cheeseburn Grange | 108 | 109 |
| Cramlington Hall | 98 | 103 |
| Darras Hall | 86 | 101 |
| Forest Hall | 68 | 62 |

750 roosting at Housesteads on 26th August was a slight increase at this locality. 1,000 were at Lower Heddon on 28th November. Flocks of over 200 were also recorded at Rochester, and Amble during the final quarter.

**Carrion Crow,** *Corvus corone*
A well-represented breeding resident.
95 at the roost at Colt Crag on 15th July was the highest count reported during the year. 33 were at Cresswell Pond on 2nd January and 37 at Widdrington tip on 22nd December.
Four pairs nested on Holy Island with one nest only three feet off the ground.
An albino bird was noted at Hauxley on 21st November and partial albinos at Jesmond Dene and Rothbury.

Hooded Crows, *C.c. cornix*, are uncommon passage and winter visitors. Singles were on the coast at Beal on 6th January and Widdrington tip on 4th and 7th February. One bird, perhaps the same as the former record, was on Holy Island on 24th April and 17th May.

In the autumn, singles were at Bradford Kaims on 19th October, Cresswell Pond on 29th November and one returned for the fifth consecutive winter to the Widdrington-Stobswood area in December.

**Raven,** *Corvus corax*
A rare resident.

A pair was noted in the Cheviot Hills between May-August and in another upland locality in October. Away from these areas, one was seen in the SW on 21st April.

**Starling,** *Sturnus vulgaris*
An abundant breeding resident and passage and winter visitor.

Winter flocks of over 1,000 were noted at Hauxley, Rayburn Lake and Druridge Bay during the early months of the year. A post-breeding roost of 15,000 birds (70% juveniles) was at Caistron on 17th July. 5,000 roosted at Blyth Harbour on 30th August and 10,000 were roosting in reeds around Holy Island Lough during October.

Three pairs nested on Coquet Island and a pair investigated the tower on Inner Farne but no breeding attempt was made.

An all-white bird was at Cramlington in April-May and two were seen on 26th September: this is now the fourth successive winter an albino has appeared at this site.

**House Sparrow,** *Passer domesticus*
An abundant breeding resident.

Post-breeding flocks were seen in July and August: 500 were at Benton and Backworth, while smaller flocks of 200 were at Big Waters and 150 at Cramlington.

**Tree Sparrow,** *Passer montanus*
A well-represented breeding resident.

The largest flocks in the first four months were 71 at Big Waters on 25th January and 60 at Meldon, near Bolam, on 23rd January. Some decline was noted with flocks of over 30 birds only being reported from one other locality (compared with 3 in 1991).

Breeding in a nestbox was reported from West Heddon and pairs were at Holywell Dene, Howick Dene and Linton Pond during the breeding season.

In the autumn and winter birds were recorded in at least ten localities with peak counts of 53 at Harlow Hill on 18th September, 51 at Ellington on 19th September and 80 at Newbiggin between 20th-31st December. Flocks of over 30 birds were also noted at Big Waters and Medburn.

**Chaffinch,** *Fringilla coelebs*
An abundant breeding species, common as a passage and winter visitor.

During the first three months there were few large flocks: 400 were at Norham in January-February, 150 were at Derwent on 1st February and 140 were at Netherwitton on 1st March.

A flock of 1,000 at Lee Hall, Wark, on 4th April were probably migrants. In April up to eight were seen on nine dates between 8th-25th on the Farne Islands.

Autumn migrants were noted at Newbiggin and the Farne Islands from 27th September, with 30 at the former site on 30th September. Up to seven birds were seen at the Farne Islands through to 8th November.

In December flocks of 150 were at Calder and Swinhoe, 110 at Denwick and 100 at Big Waters.

**Brambling,** *Fringilla montifringilla*
A common passage and winter visitor.

During the first three months records were scarce with a flock of 50 at Norham on 14th January which had declined to ten by 1st February and ten near Wooler on 16th February being the largest groups reported. A record of a female on Holy Island on 14th June was the latest spring sighting ever noted for Northumberland.

Early return in autumn was noted with one at Newbiggin on 29th August and a flock of over 20 was at Druridge Bay on 5th September.

There was a substantial arrival in late September with 80 on the Inner Farne on 27th and 50 on Holy Island on 30th. The numbers on Holy Island increased to 300 by 10th October, but numbers in the county dwindled towards the year end with 18 at Beacon Hill on 27th November and 20 at Wallington Hall on 7th December the largest flocks noted.

**Greenfinch,** *Carduelis chloris*
A common breeding resident and passage visitor.

A flock of 450 at Norham on 1st February was the only large gathering but small groups were widespread at garden feeders during the first three months.

Spring passage on the Farne Islands was very light with only four noted on three dates in early April. An adult male ringed on the Wirral, Cheshire, in November 1991 was controlled at Threestoneburn in March.

Autumn passage was not noted on the Farne Islands, but 20 in Whitley Bay Cemetery on 27th September could well have been migrants. Winter flocks included 200 at Chatton on 13th November and 150 at Whittonstall on 4th December.

**Goldfinch,** *Carduelis carduelis*
A common breeding species, well-represented on passage.

In January 50 were in the Chevington area on 4th and during March over 40 were seen at Druridge Bay Country Park on two dates.

On 10th May passage N was noted at 30-40 per hour at Bamburgh, and on the Farne Islands migrants were seen between 16th April-10th May.

Breeding was noted in many widespread localities and led to some large post-breeding flocks e.g. 84 at Stobswood on 26th September and 60 were in Elswick Cemetery, Newcastle, and 60 at Alnmouth on 19th September.

The increase in this species in recent years continued with charms of over 30 birds being noted in eight localities (compared with four in 1991) during the final quarter.

**Siskin,** *Carduelis spinus*
An abundant breeding species and passage and winter visitor.
   Few were in the major forests in the early part of the year. 200 at Big Waters and 95 at Pegswood in January were the largest flocks during the first three months.
   Ringing recoveries and controls from several sites confirmed a very rapid spring migration of birds through the county. Seven birds had travelled distances of over 200 km, including five to Highland Region, two of which were controlled there within two weeks of being ringed.
   Breeding was suspected in Jesmond Dene where birds were seen displaying in April and were present until June. The first young were seen at Threestoneburn on 31st May where over 200 were ringed during the spring. A very successful breeding season was also noted in Coquetdale and the Otterburn area.
   Flocks built up in the major forest areas in response to the moderate cone crop with 150 at Falstone on 4th August and 300 lingered in Kielder village feeding on the riverside Alders.

**Linnet,** *Carduelis cannabina*
A common breeding resident and passage visitor.
   In January 300 were on Holy Island and 1,000 at Norham but this flock declined to 500 by 1st February. 300 were at Derwent Reservoir on 9th February. Few other large flocks were noted.
   Spring passage was noted on the Farne Islands from 5th April and on 10th May passage N was noted at 40-50 per hour at Bamburgh.
   Post-breeding flocks gathered from August with 200 at Wallsend Swallow Pond on 2nd and 300 at High Hauxley on 19th September. Six other flocks of over 100 were noted in the autumn months in coastal areas. At the year end over 400 were at Whittonstall on 4th December and 200 on Holy Island on 30th December.

**Twite,** *Carduelis flavirostris*
A well-represented passage and winter visitor and an occasional breeder.
   Coastal flocks were present at four localities during the first quarter but numbers were low with 60 at Cresswell in late March the largest group. 11 were inland at Housesteads on 12th January.
   Birds were found inland at three sites suitable for breeding during April and three were still on high moorland at Coalcleugh on 4th October.
   Autumn passage was seen on the Farne Islands in early November with two on 7th and two on 8th.
   A large flock gathered in the Druridge Bay area in the final quarter with 210 recorded on 13th November. 35 were on Holy Island on 30th December.

**Redpoll,** *Carduelis flammea cabaret*
A common breeding species many of which are summer visitors, at least well-represented on passage.
   The low numbers recorded in late 1991 continued throughout the first quarter with 70 at Druridge Bay Country Park on 1st March the largest flock.
   One on 15th May was the sole spring passage record for the Farne Islands. At around the same time there were small numbers of coastal migrants noted at Hauxley and Newbiggin.

Breeding was reported at several inland localities and from the numbers reported in autumn it appears to have been a good season.

Several large flocks were recorded in the last quarter: 140 at Bolam Lake Country Park on 30th October, 70 at Colt Crag on 22nd November and 85 at Peth Foot on 20th December. Flocks of over 40 were noted at three further localities.

Mealy Redpolls, *C. f. flammea,* are an uncommon passage and winter visitor. The only record in 1992 was one ringed at Big Waters on 8th May.

**Crossbill,** *Loxia curvirostra*
A well-represented breeding species and irruptive visitor.

Poor cone crops in the major forest areas meant that very low counts of this species were in contrast to the tens of thousands in 1991. The highest count in spring was 20 at Harwood on 9th February.

There was a small influx to the Border Forests in July where the new cone crop was moderate. On the Farne Islands migrant birds were seen in September, with two on 18th and three on 21st. Numbers remained low until the year end with 15 at Stocksfield on 24th November and 12 in Dipton Wood on 27th December being the largest groups reported.

**Scarlet Rosefinch,** *Carpodacus erythrinus*
A rare passage visitor.

There were five females in spring which were the fourth-eighth spring records for the county. There were singles on the Farne Islands as follows: on Brownsman on 28th May, Inner Farne on 5th-6th June and on Brownsman on 6th June. A female was on Coquet Island on 6th June and another was at the Snook, Holy Island, on the following two days. In autumn one was on the Farne Islands on 29th September.

**Bullfinch,** *Pyrrhula pyrrhula*
A well-represented resident breeding species, uncommon on passage.

They were reported at over 40 widespread localities during the year.

The largest gatherings were in conifer forests: 50-60 at Kielder Castle and another 20 at Bakethin on 20th September and 17 in Harwood Forest on 28th December.

Successful breeding was reported in several areas during the summer.

**Hawfinch,** *Coccothraustes coccothraustes*
A rare breeding resident, uncommon on passage.

In spring numbers were low with the maximum count of 12 at Hulne Park on 2nd February and birds only reported at two other localities.

Two feeding on sunflower seeds in a Newbiggin garden on 4th April were possibly migrants as was the single flying over Holy Island village on 24th May.

The only breeding reported involved juveniles at Stamfordham on 14th August. In autumn the only records were from Hulne Park where there were five on 22nd November and up to three in December, including a first-year bird.

**Lapland Bunting,** *Calcarius lapponicus*
An uncommon passage and winter visitor.

The decline in the number of records in this species was obvious in 1992:

the only bird in spring was a single in the Druridge-Cresswell area on 1st-4th March.

In autumn single migrants were at Tynemouth on 27th September while on the Farne Islands singles were recorded on 14 days between 26th September-23rd October. There was also an elusive flock of up to seven birds on Holy Island in late October.

**Snow Bunting,** *Plectrophenax nivalis*
A well-represented passage and winter visitor.

Numbers on the coast, during the first quarter were low with 12 at Newbiggin on 15th January the maximum recorded. Inland a flock of 30 were seen at Hareshaw on 5th February. A female at Longhoughton Steel on 17th April was the only spring migrant noted.

Autumn passage was logged on the Farne Islands from 4th October-30th November, with a maximum of 14 birds. Small parties were found on the coast from 8th October with 20 at Druridge Bay on 6th November and 34 at Newbiggin on 31st December the largest flocks. Inland, 20. were at Sharperton on 24th October, 16 on Cheviot on 8th November and 13 on Hedgehope on the following day.

**Pine Bunting,** *Emberiza leucocephalos*
An extremely rare visitor.

A male found at Blakemoor Farm, Cresswell on 29th January remained until 19th February. This will either be the second or third county record as a decision is still awaited on the individual at Big Waters in February-March 1990.

The records are still under consideration by the British Birds Rarities Committee.

**Yellowhammer,** *Emberiza citrinella*
A common breeding resident.

In the first quarter there were no very large gatherings with 45 at Swallow Pond on 4th January and 80 at Elwick on 2nd February the biggest flocks reported. Flocks of over 40 were recorded at only two other localities.

Spring passage on the Farne Islands was very light, with single males reported on 6th and 9th-10th April.

Breeding in young plantations on the lowlands was confirmed at Hallington where a nest with two young was found in a small Sitka spruce on 31st July.

Frosts in December led to several large flocks developing including 250 at Medburn, 80 at Newbiggin and 60 at Big Waters.

**Ortolan Bunting,** *Emberiza hortulana*
A rare passage visitor.

An immature was on Inner Farne on 30th August and another was at Chare Ends, Holy Island on 4th October. These are the first records in the county since 1988.

**Rustic Bunting,** *Emberiza rustica*
An extremely rare visitor.

A male was on the Snook, Holy Island, on 31st May and another male was on Inner Farne on 3rd October.

These are the ninth and tenth county records. The last record was a well-watched individual on 18th-21st October 1990 at Newbiggin.

**Little Bunting,** *Emberiza pusilla*
A rare visitor.
The only record was a single at Cocklawburn dunes on 29th September.

**Yellow-breasted Bunting,** *Emberiza aureola*
An extremely rare visitor.
A female or immature was on Inner Farne between 17th-20th September. This is the sixth county record and the first since a similar bird on the Farne Islands on 1st September 1983.

**Reed Bunting,** *Emberiza schoeniclus*
A common breeding resident which can be well-represented on passage.
30 were at Newbiggin Pool on 28th March. In April, up to three passage birds were on the Farne Islands between the 5th-28th and there were three singles in May.
21 singing males were recorded at Big Waters on 21st June and ten pairs bred at Arcot Pond.
150 at Newbiggin on 27th September may have been migrants. Up to 13 were recorded on the Farne Islands from this date to 26th October. At least two migrants were in Tynemouth on 28th-30th September. In December over 50 were at Newbiggin, 30 at Prestwick Carr and 20 at Druridge Bay.

**Corn Bunting,** *Milaria calandra*
A well-represented breeding resident.
Good numbers were reported during the first quarter with 45, including a partially albino bird, at Woodhorn Pond on 28th January. In February 27 were at Norham on the 1st and 28 at Cresswell on 28th.
Although the species is declining nationally there are still reasonable numbers present with breeding birds being found in some new areas recently e.g. two singing males at N Sunderland on 8th June. During the year birds were reported at 22 coastal and near-coastal localities from St Mary's Island to Bamburgh with the bulk of the records being singing males.
A BTO survey on this species is currently investigating the full status of this species and the results will be given in a future *Birds in Northumbria*.

# CONTRIBUTORS

Allen, A; Allen, F; Almond, J M; Amies, M; Anderson, D; Anderson, M G; Annan, C R ; Armstrong, A L; Armstrong, I H; Atkinson, P; Bagshaw, E; Ball, T G; Bankier, A M; Banks, A H; Barratt, K I; Bell, D G; Bell, J G; Bell, K; Bell, M; Bellamy, P; Bentley, M; Billen, R; Birkett, J; Blair, D; Blake, T; Booth, A J; Booth, J S; Bowey, K; Bowman, A I; Bowman, G; Bradshaw, C; Brooks, J; Brooks, K V; Bruce, A; Brunt, A & L; Bull, G P Bush, P; Buskin, P A; Cadwallender, T A & M L; Calasca, H A; Carlyle, J; Carr, M J; Carr-Ellison, J; Chadwick, I; Challen, C; Charleton, P; Christer, G; Cleeves, T R; Coleman, J; Corr, A; Cosgrove, P J; Cox, N B; Crabtree, E; Craig, R; Crossan, L; Cubitt, M; Curry, A; Daggett, E; Davey, P; Davidson, I S; Davidson, P W; Davis, A J; Day, J C; Dickson, W; Dodds, G W; Douglas, I R; Dunn, R; Dutton, J; Eastlake, P B; Eccles, M; Elliot, B; Fairhurst, J; Farrar, J & M; Fergusson, J E; Fisher, I; Fiske, R M; Foggo, N; Forster, A; Forster, R S; Frankis, M P; Freeman, M A; Galloway, B; Gardiner-Medwin, D; Gilbert, P N; Gilbody, J; Gill, P; Greene, M C; Gunning, S; Hall, L; Gallowell, J; Hampton, C D; Hardy, T; Harrison, G; Harrison, J; Hart, A S; Harvey, R; Henry, M; Hepple, M; Hetherington, A G; Hewitt, A J; Hicks, R K; Hind, J; Hingston, S J; Hodgson, M K; Hodgson, M S; Holgate, D; Holmes, R M; Holt, P; Hopkin, F; Howat, J & M; Hughes, M; Hunter, R; Hutt, A; Jack, A S; Jacklin, D; James, T; Janes, A; Jardine, D C & J A; Jensen, O E; Jewitt, C; Johnson, B; Johnson, R; Johnson, W G; Johnston, A J; Jones, P R; Kanefsky, C; Kelly, W A; Kerr, I; Kerton, I; Lancaster, R & J; Lascelles, P; Law, B J; Lee, D; Lemieux, C; Lindsay, J P; Linkleter, g; Little, B; Littlefair, D; Lockwood, D; Lockwood, R J; Logan, W T; Lowther, F & M; Makin, W; McDougall, L J & A K; McKeown, D; McLevy, A D; McLoughlin, J; McNab, R; Meek, E R; Middleton, A; Moffitt, S S; Moon, G W; Mossop, A P; Mould, A D; Mowbray, A; Nattrass, M; Nattress, K; Norman, G; Norman, R; Norris, E T & J R; Oddie, W E; Osborne, N F; Parnaby, S P; Parrack, J D; Phillimore, A B; Pickard, D; Plenty, M D; Priest, A; Purvis, A; Quarterman, H; Redfearn, C P F; Redgrave, K; Regan, K W; Rhodes, D; Richards, A J; Richardson, B G; Richardson, D C; Richardson, J; Richardson, M; Robinson M A; Robson, K; Robson, L; Robson, S; Rodgers, K; Rossiter, B N & A F; Ruddock, W M; Rushbrooke, G S; Russell L & K; Rutter, J W; Savage, W G; Scott, M; Scott, R J; Sexton, S; Shannon, D R; Sharp, M J; Shaw, M B; Slack, E; Smart, D; Smith, M R; Smith, S B; Snart, S; Sneyd, J E; Sorrie, G A; Starling, A E; Steele, E J; Steele, J G; Stephenson, R K; Stimpson, A; Stott, J; Tams, T J; Taylor, G; Telfer, N; Thain, T; Thomas, M; Thomas, W I; Thompson, B; Thomson, M; Thornton, M; Tidmarsh, J; Tilmouth, A; Tindle, H; Todd, A L; Todd, J R; Tucker, K; Turner, D M; Vaughan, R & M; Veitch, P; Votier, S C; Walton, J; Walton, K; Walton, R; Watson, A M; Watson, D; Watson, P M; Watson, T P M; Waugh, R G; Welbury, J; Wentzel, J; West, P W; Westerberg, S; Winter, M; Wiseman, N; Wright, S; York, K W; Bamburgh Ringing Station; Birdline North East; Druridge Bay Nature Reserve; English Nature; Hauxley Ringing Station; National Trust; Prior's Park Ringing Station; Royal Society for the Protection of Birds; Seaton Sluice Watch Tower.

# ADDITIONS AND CORRECTIONS TO EARLIER REPORTS

## ADDITIONS

*Birds in Northumbria 1990*

Wigeon, *Anas penelope*
    Caistron, July
    A pair fledged four young

*Birds in Northumbria 1991*

Red-necked Grebe, *Podiceps grisegena*
    Lough in SW, June-July
    One bird summered with some indications of territorial behaviour

Red Kite, *Milvus milvus*
    Belford, 9th August
    One Scottish wing-tagged bird

Lesser Spotted Woodpecker, *Dendrocopos minor*
    Riding Mill, 19th October
    Single bird

Stonechat, *Saxicola torquata*
    S of Wark Forest, June
    Two pairs present

Pied Flycatcher, *Ficedula hypoleuca*
Blue Tit, *Parus caeruleus*
Great Tit, *Parus major*
Nuthatch, *Sitta europaea*
    Nest box data for Wallington Hall for years 1989-91 is printed in bulletin for August 1992 at page 131

Carrion Crow, *Corvus corone corone*
    Kilham, 23rd June
    Two newly-fledged juveniles were seen showing characteristics of Hooded Crow, *C. c. cornix*. The parents were both typical Carrion Crows.

# CORRECTIONS

*Birds in Northumbria 1990*

Richard's Pipit, *Anthus novaeseelandiae*
    Hauxley, 23rd September
        On review by the County Records Committee, this record was found to be no longer acceptable.

*Birds in Northumbria 1991*

Razorbill, *Alca torda*
    Farne Islands
    110 pairs bred

Blue Tit, *Parus caeruleus*
    Fontburn
    81 young were fledged

# BBRC DECISIONS

The following records were described in earlier reports as being under consideration by the British Birds Rarities Committee. This committee has now given the following decisions:

*Birds in Northumbria 1986*

White-billed Diver, *Gavia adamsii*
    Holy Island, 23rd February         Not accepted
        Originally accepted at the county level, this record is currently under review by the County Records Committee.

*Birds in Northumbria 1987*

Bonelli's Warbler, *Phylloscopus bonelli*
    Howick, 24th August         Not accepted
        Originally accepted at the county level, this record has been reviewed by the County Records Committee who agreed to concur with the BBRC view.

Arctic Redpoll, *Carduelis hornemanni*
    Big Waters, 6th January         No conclusion possible
        Unfortunately, the slides were lost and some measurements were not taken and it was not possible for BBRC to substantiate the record against the latest identification criteria. The County Records Committee, who earlier had accepted the record, also consider the record to be not fully substantiated.

*Birds in Northumbria 1989*

Little Shearwater, *Puffinus assimilis*
    Seaton Sluice, 17th July                                   Accepted
    Annstead, 10th September                         Not accepted
    The County Records Committee, which earlier had made no definite decisions, followed the views of BBRC in accepting the Seaton Sluice record and not accepting the Annstead sighting.

Black Scoter, *Melanitta nigra americana*
    Buston Links, 19th January                        Not accepted
    Originally accepted at the county level, this record is currently under review by the County Records Committee.

Gull-billed Tern, *Gelochelidon nilotica*
    Cresswell, 1st July                                       Accepted

*Birds in Northumbria 1990*

Siberian Stonechat, *Saxicola torquata maura/stejnegeri*
    Bamburgh, 21st October                               Accepted

*Birds in Northumbria 1991*

American Wigeon, *Anas americana*
    Cresswell, 22nd-29th December                 Accepted

Ring-necked Duck, *Aythya collarus*
    all records                                                 Accepted

Black Kite, *Milvus migrans*
    Kielder, 18th April                                       Accepted

Pacific Golden Plover, *Pluvialis fulva*
    Druridge Pools, 22nd-23rd June                  Accepted

Terek Sandpiper, *Xenus cinereus*
    Blyth, to 5th January                                  Accepted

Lesser Crested Tern, *Sterna bengalensis*
    Farne Islands, 14th May-18th August          Accepted
    Hauxley, 16th June          Accepted as definitely of this species.
    The County Records Committee had accepted the Hauxley record as 'large yellow-billed tern probably of this species'. After a review of the record, the County Records Committee concur with the observer that specific identification was not possible. BBRC are reviewing their position on this type of record.

Red-throated Pipit, *Anthus cervinus*
    Farne Islands, 29th May                               Accepted

Citrine Wagtail, *Motacilla citreola*
    Hauxley, 15th-16th May                                           Accepted

Subalpine Warbler, *Sylvia cantillans*
    Farne Islands, 26th April                                        Accepted

Greenish Warbler, *Phylloscopus trochiloides*
    Farne Islands, 22nd August                                   Accepted

Radde's Warbler, *Phylloscopus schwarzi*
    Tynemouth, 11th-14th October                             Accepted

Dusky Warbler, *Phylloscopus fuscatus*
    Farne Islands, 27th October                                   Accepted

Parrot Crossbill, *Loxia pytyopsittacus*
    Kielder Forest, 19th January                                  Accepted
    Kielder Forest, 24th February                               Accepted
    The second record was submitted by the observer directly to BBRC who considered it to be the same individual as the one seen on 19th January. The County Records Committee also accepted the February record but considered this bird to be a different individual to that seen earlier.

Little Bunting, *Emberiza pusilla*
    Farne Islands, 25th-26th September                 Accepted
    Farne Islands, 10th October                                 Accepted
    Newbiggin, 10th October (3 birds)                    Accepted
    Farne Islands, 12th-14th October                        Accepted

## DECISIONS PENDING

The following records, from 1990 or earlier, are still under consideration by the British Birds Rarities Committee. No decision has been taken at the county level on those for Sooty Tern, Pine Bunting and Two-barred Crossbill.

Little Shearwater, *Puffinus assimilis*
    Tynemouth, 18th August 1986

American Golden Plover, *Pluvialis dominica*
    St Mary's Island, 13th October 1987

Bridled Tern, *Sterna anaethetus*
    Cresswell, 31st July 1988

Sooty Tern, *Sterna fuscata*
    Long Nanny Burn, 10th-11th and 20th July 1988
    Farne Islands, 8th-12th July 1988

Dusky Warbler, *Phylloscopus fuscatus*
    Farne Islands, 28th October-7th November 1989

Siberian Stonechat, *Saxicola torquata maura/stejnegeri*
    Newbiggin, 20th October 1990

Pine Bunting, *Emberiza leucocephalos*
    Big Waters, 17th February-16th March 1990

Two-barred Crossbill, *Loxia leucoptera*
    Harwood Forest, 24th December 1990-16th March 1991

## EXOTICA

The following records refer to birds most likely to have had captive origin. The value of publishing such records is to monitor possible colonisations by exotics, that is species not on the British list.

Chilean Flamingo, *Phoenicapterus chilensis*
| | |
|---|---|
| Druridge, 10th April | two |
| Coquet Island, 25th June | two |
| Newbiggin, 23rd July | two |

Cinnamon Teal, *Anas cynaoptera*
| | |
|---|---|
| Hauxley, 14th May | two |

Sulphur-crested Cockatoo, *Cacatua galerita*
| | |
|---|---|
| Birling Carrs, 22nd November | one, obviously an escape |

# THE NIGHTJAR BREEDING SURVEY 1992
## by Tom Cadwallender and Colin Jewitt

Early this century, Bolam (1912) described the Nightjar, *Caprimulgus europaeus*, as occurring sparingly in the region while earlier Chapman (1889) had considered that it was probably more common than it appeared because of its nocturnal habits.

Following these comments from the two eminent ornithologists very little seems to have been recorded, save the idea that a steady decline was taking place. This appeared to have become widespread by the mid-1930s and accelerated by the 1950s when it was noted right across NW Europe. Probable causes included habitat loss, disturbance, climatic changes and a decrease in prey due to an ever-increasing use of pesticides (Cramp 1985).

Surveys co-ordinated by the British Trust for Ornithology (BTO) in 1957 and 1981 (Macfarlane 1981), and observations during the 1968-73 breeding atlas work, confirmed no more than eight-ten churring males in Northumberland, all in traditional sites, such as moorland edge, in the N and SW.

In 1992 the BTO and the Royal Society for the Protection of Birds (RSPB) organised another national survey, using mainly amateur ornithologists to cover the vast majority of sites. The Forestry Commission and private forestry owners provided valuable assistance with stock maps and encouraged their employees to help in more remote areas.

Northumberland's survey was based on the following criteria-

* Traditional sites
* Sites found during survey work in 1989-1992 for the new breeding atlas.
* Potential sites resulting from a study of forestry maps indicating suitable ages of trees.

**Habits and Habitat**:

From early May, the Night Hawk, Moth Hawk, Fern Hawk, Goatsucker or Gabble Ratchet, to give the Nightjar just a few of its folk names, arrives from African wintering grounds. It migrates nocturnally, singly or in small groups, males usually arriving first. Occasionally birds arrive already paired and consequently in low density areas do not need to sing continually to attract a mate or hold territory so some can be easily missed.

Habitat is essentially on open ground but sometimes in clearings or along the edge of mature conifer or deciduous woodland. In Northumberland the favoured habitats are cleared or replanted areas with a rich undergrowth of Heather, *Calluna/erica sp.*, Bilberry, *Vaccinium myrtillus*, and Bracken, *Pteridium aquilinum*.

Some of the highest densities are on Forestry Commission land with unburned branches and tops from the last crop. Peak numbers are in felled and replanted areas of 50-250 acres with forest remaining on one or more sides, ranging from seedlings to trees of up to four metres. The birds will use rides as territory boundaries. The favoured areas are typically on high ground but with easy access to lower localities with a rich diversity of fields, hedges, valleys and sheltered wood edges where birds can hunt for flying insects, mainly the larger species of *Lepidoptera* and *Coleoptera* (Cramp 1985).

**NIGHTJAR – many more than previously suspected were found**

often not available at higher altitude. The recent increase in suitable habitat from forestry and the current run of warm dry summers has possibly resulted in an increase in young being raised and consequently a rise in the population.

### Northumberland Status:

The 1992 survey revealed 43-45 churring males or territorial pairs. This very substantial rise on previous figures was reflected across the country with Northumberland showing one of the highest apparent increases. However, it is not clear whether this was due to the factors mentioned above or because of the involvement of more observers covering huge areas of potentially suitable habitat, or a combination of both.

In the past the Tyne Valley and SW have been the Nightjars strongholds. The 1992 survey showed that this general picture was unchanged with these areas holding 25 churring birds, 58% of those found. Hexhamshire with 18 singing males had by far the greatest concentration.

Elsewhere the population was widespread but thin. Pockets of one-five males were found in upper and middle Coquetdale, Thrunton Woods, Kyloe and Holburn Moss. However none were found in Kielder Forest and many other apparently suitable areas produced negative results. However, these sites, in time, could provide Northumberland with a larger population than was discovered in 1992.

### Acknowledgements:

The 1992 survey could not have been completed without a very dedicated band of observers and the assistance of the Forestry Commission, in particular Clive Large at Rothbury and David Jardine and James Ogilvie at Kielder. Peter Hale of Hale Associates supplied maps and gave access.

### Observers:

N. Anderson, G. Bowman, M & T. Cadwallender, C. Challen, L. Crossen, I. Davidson, K. Dawson, W. Dickson, I. Douglas, J. Dutton, M. Frankis, B. Galloway, M. Henry, M. Holmes, A. Janes, D. Jardine, C. Jewitt, I. Kerr, G. Linkleter, D. Littlefair, W. Moon, A. Mossop, M. Richardson, B. N. Rossiter, W. Ruddock, J. Steele, A. Stimpson, A. Tilmouth, P. & E. Taylor, T. Watson and S. Winter.

*Apologies for any accidental ommissions.*

### References:

Chapman, A. (1889) *Bird-Life of the Borders.*
Bolam, G. (1912) *The Birds of Northumberland and the Eastern Borders*
Macfarlane, L. G. (1981) Nightjar Survey 1981, *Birds in Northumbria 1981*
Cramp, S. ed. (1985) *The Birds of the Western Palearctic*, Vol. IV.

# BREEDING SURVEY OF SHELDUCK IN NORTHUMBRIA 1992
## by Mike S. Hodgson

During the spring and summer of 1992, following two pilot surveys, the Wildfowl and Wetlands Trust organised a national survey of breeding Shelduck, *Tadorna tadorna*. The breeding population of Britain has never been accurately determined although the National Waterfowl Counts index for the species has gradually risen from the 1960s to the 1980s. The most recent estimate of the current British breeding population is 'probably more than 15,000 pairs' (Delany 1992).

In Northumberland the Shelduck is described as a well-represented passage and winter visitor and an uncommon breeding species. Galloway and Meek (1977) stated that as a breeding species it appeared to be most abundant about 1947 when over 100 pairs nested on or near Lindisfarne but following a considerable decline only about 30 pairs breed in the county with their southern limit near Cresswell.

The 1992 survey was conducted in two parts with straightforward instructions being provided for observers. The first counts (up to three) could be carried out in the period 25th April-17th May, ideally within two hours of low tide and on an afternoon or evening in the first ten days of May. The total number of birds seen was the most important count and observers were asked to sub-divide that total into territorial males, number of pairs, number in non-breeding groups and, if time permitted, the number of pairs showing territorial behaviour. The second counts (ideally two) were to be carried out in the period 27th June-2nd August, preferably during the week-end of 11th-12th July. The counts required were the total number of adults and the total number of young seen with the latter split into creche and, where possible, brood sizes.

The survey was well supported with observers from the Northumberland and Tyneside Bird Club, the National Trust and RSPB providing full coverage along the entire coastline from Berwick Harbour to the Blyth Estuary and at one inland site in the Kyloe Hills. It was felt that any other breeding records away from these areas could be monitored from the monthly record cards submitted to the County Recorder.

The full results of the survey are shown in Table 1. These have been adjusted slightly for the summer counts where young were seen before, but not subsequent to, the count dates and for one inland site at Caistron.

The spring counts show that at least 74 territorial pairs were present although less than half of these appear to have bred successfully (i.e. hatched at least one young). The total of 230 young noted during the second part of the survey represents about 34 broods from 13 localities. This compares favourably with the last county survey in 1979 which located 27 broods (Heavisides and Hodgson, 1980). It would appear that Shelduck have never been more than occasional breeders away from the coast and the site at Caistron, where breeding first took place in 1989, is the furthest inland to be recorded. The site at Castle Island, where breeding also took place in 1990, represents a slight southward range extension. It is interesting to note that about 82% of broods were located in sites which were either reserves or wardened in some way during the breeding season.

**Table 1. Shelduck Survey Results 1992**

| Locality | Total No. of Birds | SPRING COUNTS No. of Territorial Males | No. of Pairs | No. in Non-breeding Groups | No. of Pairs Territorial | SUMMER COUNTS Total No. of Adults | Total No. of Young |
|---|---|---|---|---|---|---|---|
| Berwick to Cheswick | 0 | | | | | 0 | 0 |
| Holy Island & Fenham Flats | 91 | 5 | 25 | 36 | 30 | 16 | 15 |
| Guile Point to Budle Bay | 89 | 4 | 17 | 50 | 15 | 75 | 5 |
| Bamburgh to Beadnell | 3 | 0 | 0 | 3 | 0 | 0 | 0 |
| Farne Islands | 5 | | 2 | | 2 | 4 | 15 |
| Beadnell to Craster | 12 | 5 | 5 | 2 | 3 | 6 | 25 |
| Craster to Alnmouth | 22 | | 11 | | 2 | 8 | 7 |
| Aln Estuary to Birling Carrs | 18 | | 9 | | 0 | 0 | 0 |
| Coquet Estuary | 13 | 5 | 5 | 3 | 5 | 2 | 18 |
| Coquet Island | 18 | 4 | 8 | 10 | 4 | 3 | 12 |
| Hauxley Nature Reserve | 30 | | | | 7 | 7 | 36 |
| Druridge Bay C.P. | 10 | | 5 | | 4 | 4 | 20 |
| Cresswell Pond | 9 | | | | 3 | 3 | 9 |
| Warkworth Lane Pond | 2 | | | | 1 | | |
| Linton Pond | 5 | | | 5 | | | |
| Wansbeck Estuary | 20 | 7 | 7 | 6 | 6 | 6 | 35 |
| Blyth Estuary | 12 | | 6 | | 5 | 0 | 0 |
| Holburn Moss/Kyloe Hills | 6 | | 3 | | 1 | 0 | 0 |
| Caistron | 8 | | 4 | | | | 24 |

I would like to express my thanks to all of the observers who took part in the survey and made it so successful.

**List of Contributors:**
M. L. Cadwallender, T. A. Cadwallender, T. R. Cleeves, J. C. Day, A. Forster, M. P. Frankis, M. Freeman, M. S. Hodgson, P. Lascelles, R. J. Lockwood, L. J. McDougall, D. McKeown, D. C. Richardson, B. N. Rossiter, A. Tilmouth, J. Walton, J. A. Wardropper, T. Watson, National Trust, RSPB
*With apologies for any accidental omissions*

**References:**
Delany, S. *Pilot Survey of Breeding Shelduck in Great Britain and Northern Ireland 1990-1991*
Galloway, B. & Meek, E. R. *Northumberland's Birds*
Heavisides, A. & Hodgson, M. S. *Birds in Northumbria 1979*

# NEW SPECIES FOR THE COUNTY

## PIED-BILLED GREBE AT DRURIDGE POOLS
### by Ian Fisher

Saturday, 26th December 1992 was foggy and overcast. After meeting Michael Carr we set off up the coast to 'do the bay'. There was very little at Hauxley Reserve so we drove to Druridge Pools, arriving about 14.00 hours. On opening the shutters of the Oddie Hide, overlooking the main pool, I noticed a medium-sized grebe with a longish neck and a large angular head just as it dived about ten metres from the shore.

I jokingly remarked that it looked like a Pied-billed Grebe, *Podilymbus podiceps*, and proceeded to scan the pool. The bird reapeared a few seconds later, having caught a fish. Its jizz again struck me as being unlike that of Little Grebe, *Tachybaptus ruficollis*, and this time it was examined with binoculars. I drew MC's attention to it and a quick examination of its head pattern and bill shape confirmed that we were indeed watching a Pied-billed Grebe. I cursed the fact that it was the first time in months I'd come out birdwatching without my camera!

We watched the bird and took notes for about 15 minutes and then left to spread the news and give everyone a late Christmas treat. Unfortunately, most of those we telephoned were either away, already out birdwatching or were too full of Christmas cheer and only seven managed to make their way to the pond that afternoon. However, the grebe remained well into January 1993, giving many hundreds of observers the opportunity to watch it at leisure.

Twice on 26th when the bird caught sight of us in the hide, it sank vertically and then reappeared some distance away. On 27th, while moving across the pond, it dived and re-emerged with only its head showing and proceeded to swim this way before resurfacing completely.

The bird's behaviour was very interesting, alternating between intensive periods of feeding and resting, sometimes in the shelter of sparse areas of vegetation. When feeding it was usually very active, often swimming at high speed across the pool. When diving it had a very impressive strike rate at catching small fish, mainly Sticklebacks, *Gasterosteus aculeatus*, which it appeared to pull from weed which was also dredged up. On one occasion from 11 dives it caught ten fish. It also fed frequently on Common Frogs, *Rana temporia*, which it would bring to the surface. Often it would take five to ten minutes to kill a frog but on occasions was seen to swallow them whole, even though they were large.

The grebe was also very aggressive to Goldeneye, *Bucephala clangula*, and on several occasions when feeding would suddenly charge across the surface towards them forcing them into flight.

The following description was taken during several hours observation between 26th-28th December:-

**Size**: A medium-sized grebe about that of Slavonian Grebe, *Podiceps auritus*.

**Jizz**: The large angular head, 'chicken' bill, long thickish neck combined with the relatively long rectangular body, squared off rear end and spikey tail, gave the bird a very distinctive shape even at distance.

**Bare parts**: Bill: short and stocky and lacking the 'gape' of Little Grebe, dull yellow with a faint dark subterminal mark on the upper mandible. Eyes: dark. Legs and feet: grey.

**Upperparts**: Crown and nape: dark brown going above and not through the eye, extending down hindneck and on to mantle and tail etc. which were also dark brown. The underwings were not seen.

**Underparts**: Chin and throat: white. Rest of face: buff brown with a paler eye ring. Foreneck: rufous brown extending to breast. Breast: greyish brown. Belly: white. Flanks: buff brown becoming darker towards belly. Undertail coverts: white.

This was the first record of this North American species for Northumberland and only the 16th for Britain and Ireland. It has been accepted by the County Records Committee and, at the time of writing, is still under consideration by the British Birds Rarities Committee.

# SEMIPALMATED SANDPIPER ON THE FARNE ISLANDS

## by Peter Bush

After an excellent period of wind from the E, the airstream moved to the N and W on 15th June 1992. Not really expecting to find anything special I visited Knoxes Reef, which can only be reached from Inner Farne by inflatable, to check my study area of Fulmars, *Fulmarus glacialis*.

While watching the sitting Fulmars I was distracted by a Ringed Plover, *Charadrius hiaticula*, calling nearby. I located the bird feeding among decaying seaweed with a small group of Turnstones, *Arenaria interpres*, Dunlin, *Calidris alpina*, and a small 'stint'.

Familiar with Little Stints, *Calidris minuta*, in most plumages, I immediately realised this was something interesting! Trying not to panic, I took brief notes before a hurried trip back to the Inner Farne to fetch Bill Makin, Andy Robinson, Paul Allen and telescopes.

We spent the next hour watching the bird and taking notes as the group remained feeding along the tideline, moving closer as the water rose. From discussion, the four of us came to the conclusion that we were looking at either a Western Sandpiper, *Calidris mauri*, or more probably a Semipalmated Sandpiper, *Calidris pusilla*!

Unfortunately, the first boatload of visitors was seen approaching Inner Farne, open only in afternoons in the breeding season, so we made a hurried departure leaving Bill Makin to continue observations.

While manning the information centre and trying to consult available literature, an excited radio call from Bill confirmed that he had managed to see partial webbing between the bird's toes as it stood facing him on a rock less than seven metres away. From this distance he had also been able to confidently separate the bird from Western Sandpiper.

Trying to contain my excitement, I had to spend the next three hours on normal duties, answering the questions of visitors with the knowledge that I had found Northumberland's first Semipalmated Sandpiper. After the departure of the last visitors we returned to the reef and were able to sit and watch the bird and take a lengthy description at leisure.

Unfortunately, due to the sensitive nature of Knoxes Reef with its nesting Fulmars and Ringed Plovers as well as Eiders, *Somateria mollissima*, and Oystercatchers, *Haematopus ostralegus*, and its inaccessibility, we had to make the reluctant decision not to release the news until the bird departed.

The bird remained for three days during which we spent long hours watching it and wasting a lot of film! The bird had departed by the morning of 19th June.

The description has been accepted by the County Records Committee and at the time of writing is still under consideration by the British Birds Rarities Committee.

## SWINHOE'S STORM-PETRELS AT TYNEMOUTH
### by Mark Cubitt

Since July 1989 the five records of three individuals of Swinhoe's Storm-petrels, *Oceanodroma monorhis*, at Tynemouth pier have interested bird-watchers locally, nationally and internationally. The protracted nature of their identification and the fact that the only British records have been at Tynemouth, despite efforts elsewhere, have both contributed to the intrigue.

The records have all occurred after midnight on the following dates: 23rd and 26th July 1989, 6th July 1990, 31st July 1991 and 30th July 1992. All were caught during the tape-luring of Storm Petrels, *Hydrobates pelagicus*, on the beach and were attracted by recordings of the burrow call of that species. The latter three records all involved the same individual which still bears the ring given in 1990.

It took over three years to positively identify them as Swinhoe's Storm-petrels. We split the research into three main areas. The first area was plumage and measurements upon which we based our first published note shortly after the first capture. The note stated that they were 'Swinhoe's Storm-Petrel or a closely related sub-species'. It was the latter possibility that forced us into looking into vocalisation and DNA sequencing which have been the other main areas of research.

Our birds have a very similar structure to Leach's Petrel, *Oceanodroma leucorhoa*, which is the British representative of the *oceanodroma* family. They have some minor size differences when compared to the large races of Leach's, for example the British population, such as bill length and tarsus length being greater and tail length being shorter in proportion to the size of the bird.

The two most significant differences are the all-dark rump and the white showing on the outer six primary feather shafts of Swinhoe's. Some races of Leach's do have dark rumps, but these races are much smaller than Swinhoe's. Individuals of the British population can have significant variations in rump coloration but will always have some white present. The presence of significant white primary shafts is restricted within this family to Swinhoe's and Matsudaira's Petrels, *Oceanodroma matsudairae*, the latter being much larger.

The second area of research, which has taken longest, is the analysis of DNA sequences. Through this we have discovered that species within the family, such as Leach's and Swinhoe's, are quite distantly related to each other in evolutionary terms even though they look very similar. The analysis also showed that our birds had identical sequences to those of Swinhoe's from Korea and Russia in the part of the gene that is looked at in this type of research.

The reason why they are genetically so different yet look so similar may be because there is little pressure on them for evolutionary change in that they occupy a similar ecological niche and almost all are nocturnal during their breeding acitivites and so plumage differences are of little value. Research shows that it is likely that it is the vocalisations that provide an important mechanism for birds to identify the species and sex of another individual in the darkness of a colony.

This was the third area of our research. Fortunately, the Japanese have recently been researching the vocalisations of Swinhoe's and so we were able to get access to their recordings which were not available in the United Kingdom. Their research shows significant differences between the vocalisation of Swinhoe's and Leach's and it allowed us to match our recordings of one of our birds to those of Swinhoe's.

The Tynemouth records have been submitted to the British Birds Rarities Committee which will pass them on to the British Ornithologists' Union Records Committee when they have completed their circulation.

As the two ringers carrying out the Tynemouth operations, Mary Carruthers and I would like to thank all those from the club who have come down to support us. Adam Hutt, Les Hall and David Hirst deserve special thanks for their regular assistance.